Erwin Schrödinger

Collected Papers
On Wave Mechanics

With an Introduction by Valia Allori

MINKOWSKI
Institute Press

Erwin Schrödinger
12 August 1887 – 4 January 1961

ISBN: 9978-1-927763-80-3 (softcover)
ISBN: 978-1-927763-81-0 (ebook)

Minkowski Institute Press
Montreal, Quebec, Canada
http://minkowskiinstitute.org/mip/

For information on all Minkowski Institute Press publications
visit our website at http://minkowskiinstitute.org/mip/books/

CONTENTS

ii

NOTE TO THE NEW PUBLICATION

This is a new publication of Erwin Schrödinger's book *Collected Papers On Wave Mechanics*, translated from the second German edition (Blacki and Son Limited, London and Glasgow 1928).

Schrödinger's book was typeset in LaTeX and noticed typos in the texts and the equations were corrected.

19 May 2020   Minkowski Institute Press

INTRODUCTION

Valia Allori
Philosophy Department
Northern Illinois University

This is a collection of six papers that Schrödinger published at the rate of almost one a month in 1926. Three more papers written in 1927 were added to the second German edition of the book published in 1928, before being translated into English.

This book contains the foundation of wave mechanics as a theory of matter, in which the now-famous Schrödinger equation first appears. As it is acknowledged by the author himself in his introduction, Schrodinger wrote the first paper without knowing exactly what deeper implications it may have. It was like exploring a dark, unknown room with a flashlight: you never know whether there is actually a door until you find one. Schrödinger's inquiry was driven by the desire of understanding what lies behind the phenomena. He believed that scientific theories should be taken as describing the mechanisms which give rise to the experimental results rather than merely systematizing them. In other words, he was a scientific realist searching for intuitive models which he believed would shed some light on the nature of things, just as a flashlight would indicate the door in the room. In his first paper in this collection, "Quantisation as a Problem of Proper Values (Part I)," his aim is to provide a deeper explanations of the 'quantum rules' which were merely postulated to reproduce the data. This explanation is tentative, but it suggests that there may be some oscillatory phenomenon in the atom, so that one can try to build a 'wave mechanics' to account for the experimental data. The other papers continue in this enterprise, as they provide the building blocks of what Schrödinger regarded as a promising model to describe reality, refining the ideas, and filling in the holes. Ultimately, however, the project turned out to be unsuccessful. In fact while in the papers in this collection Schrödinger points out at the difficulties of his project and he is hopeful of solving them, the situation quickly changed after 1928. For a variety of reasons[1] the Copenhagen School and its anti-realist attitude had won over most physicists, and Schrödinger stopped working on quantum theory if not to criticize it. So, while one may naturally think of the first paper in this collection as a starting point, namely the birth of the new quantum theory based on waves, I think this can also be seen as a point of arrival, as Schrödinger did little more work on quantum theory after these papers.[2]

In any case, Schrödinger's intuitive model of the world is a mechanics of waves, and it is interesting to see where it is coming from. By a quick look at Schrödinger's work before this set of papers, one notices several examples of his desire to understand things intuitively. First of all, as Schrödinger himself remarked multiple times,[3] he was strongly influenced by Boltzmann (one of his teachers at the University of Vienna was Boltzmann's student) and his realist attitude of

---

[1]See e.g. M. Beller (1999): *Quantum Dialogue: The Making of a Revolution*, University of Chicago Press, for more on this.

[2]See also L. Wessels (1979): "Schrödinger's Route to Wave Mechanics," *Studies in History and Philosophy of Science Part A* 10 (4): 311-340 and references therein.

[3]P. A. Hanle (1975): *Erwin Schrödinger's Statistical Mechanics, 1912-1925*. PhD Dissertation, Stanford University. This book examines in detail the influence of the Boltzmann statistical tradition on Schrödinger.

understanding the macroscopic phenomena in terms of the microscopic classical dynamics. For instance, just like Boltzmann, at some point Schrödinger believed that the atoms, the fundamental particles postulated in classical mechanics, actually exist, and in 1914 he even wrote a paper to support the atomic hypothesis using an analysis in terms of elastic phenomena.[4] Also, to support that he was a realist, let me notice that in 1925 Schrödinger wrote a short philosophical essay arguing against Mach's view that economy and efficiency are the only factors to consider in scientific investigations.[5] So Schrödinger was a realist, but why about waves? Part of the reason can be tracked to the work of Bohr, Kramers and Salter[6] who in 1924 proposed a theory of radiation emission in terms of 'virtual fields.' Schrödinger was enthusiastic about the paper because it provided a way of physically visualize what was happening in radiation emission phenomena, but he did not understand why the fields needed to be virtual. Consequently he reformulated the theory in terms of a real field instead[7] along the lines of de Broglie, who complemented Einstein's idea that to every wave there is a particle with the

---

[4]E. Schrödinger (1914): "Zur Dynamik elastisch gekoppelter Punktsysteme," *Annalen der Physik* 4 (44): 914-934.

[5]This essay, 'Seek for the Road,' has been published later in E. Schrödinger (1964): *My View of the World*, Cambridge: Cambridge University Press: 3-60.

[6]N. Bohr, H.A. Kramers, J.C. Slater (1924): "Über die Quantentheorie der Strahlung," *Zeitschrift für Physik* 24 (1): 69–87. Available also in English, N. Bohr, H.A. Kramers, J.C. Slater (1924): "The quantum theory of radiation." *The London, Edinburgh, and Dublin Philosophical Magazine and Journal of Science* 47 (281): 785–802

[7]L. de Broglie (1923a): "Ondes et quanta," *Comptes rendus de l'Académie des Sciences* 177: 507-510; L. de Broglie (1923b): "Quanta de lumière, diffraction et interférences," *Comptes rendus de l'Académie des Sciences* 177: 548-551; L. de Broglie (1923c): "Les quanta, la théorie cinétique des gaz et le principe de Fermat," *Comptes rendus de l'Académie des Sciences* 177: 630-632.

idea that to every particle there is a wave.[8]

Another piece of the puzzle to understand how he arrived at wave mechanics comes from his reflection about Boltzmann's statistical mechanics. In 1925 he wrote three papers of his on quantum gases.[9] He wanted to give an explanation of Planck's radiation law in terms of a gas of light quanta and by applying the Bose-Einstein statistic developed the year before.[10] Schrödinger was unsatisfied with the fact that the particles in a Bose-Einstein gas are counted in a way which he regarded as physically artificial, and he thought that something in their nature must explain why. However instead of providing a microscopic description of the gas, Schrödinger proposed to treat the gas as a whole, as a matter of convenience. In particular in the third gas paper he used de Broglie's theory to describe the gas not as a collection of particles but as a matter wave. By doing this he showed that the Bose-Einstein counting procedure can be understood as a counting method for standing wave modes. Schrödinger saw this as a strong indication that there had to be something right about the matter wave hypothesis. In other words, while originally the idea of describing the gas as a matter wave was simply a useful tool to overcome the lack of understating of the nature of the gas particles, now it is seen as providing some insight on how to get such understanding:

---

[8]E. Schrödinger (1924): "Bohrs neue Strahlungshypothese und der Energiesatz," *Die Naturwissenschaften* 12: 720-724.

[9]E. Schrödinger (1925): "Bemerkungen über die statistiche Entropiedefinition beim idealen Gas," *Sitzungsberichte der Preußischen Akademie der Wissenschaften. Physikalisch-mathematische Klasse*: 434-441; E. Schrödinger (1926): "Die Energiestufen des idealen einatomigen Gasmodells," *Sitzungsberichte der Preußischen Akademie der Wissenschaften. Physikalisch-mathematische Klasse*: 23-36; E. Schrödinger (1926): "Zur Einsteinschen Gastheorie," *Physikalische Zeitschrift* 27: 95-101.

[10]A. Einstein (1924): "Quantentheorie des einatomigen idealen Gases," *Königliche Preußische Akademie der Wissenschaften. Sitzungsberichte*: 261–267.

"particles are nothing more than a kind of 'wave crest' on a background of waves."[11] Schrödinger therefore already in this paper entertained the possibility that particles could be reduced to localized wave packets, even if he immediately realized that such a packet would quickly spread out. Be that as it may, after these papers Schrödinger wanted to see whether he could understand other experimental findings in terms of oscillatory phenomena, and started with the spectrum of the hydrogen atom. He wanted to identify the orbits as suitable modes of vibrations of a wave equation. In other words, he wanted to reproduce the energy levels of the hydrogen atom as the proper values (*eigenvalues*, in modern terminology) of a suitable wave equation (hence the title of the first paper of this collection). Interestingly, since he wanted to have the new theory compatible with relativity, in December 1925 he found a relativistic wave equation, now known as the Klein-Gordon equation. However, he soon discarded it because it gave the wrong results.[12] This was the first attempt at wave mechanics, before publishing the papers in this volume.

Before leaving you to enjoy the papers for yourselves, let me point out some of their salient ingredients, together with some more history along the way. In January 1926 Schrödinger wrote the first paper in this volume namely "Quantisation as a Proper Value Problem (Part I)," in which as we just mentioned he lays the foundations of wave mechanics. Given that the relativistic equation was not empirically adequate, Schrödinger develops a non-relativistic ver-

---

[11]"Korpuskel, nach welcher dieselbe nichts weiter als eine Art 'Schaumkamm' auf einer den Weltgrund bildenden Wellenstrahlung ist." E. Schrödinger (1926): "Zur Einsteinschen Gastheorie," *Physikalische Zeitschrift* 27: 95.

[12]He mentions the fact that the relativistic equation would lead to half-integral azimuthal quantum numbers in the first paper. A proposed relativistic equation for a single electron can be found in "Quantisation as a Proper Value Problem (Part IV)" paragraph 6. See later for more on this.

sion which instead could account for the observed values of the hydrogen spectrum. The aim of this paper is straightforward: show that the hydrogen spectrum can be reproduced in terms of nodes of a suitably vibrating string, along the lines of de Broglie's hypothesis of matter waves. However, the presentation is extremely abstract: there is little mention of his metaphysical hypothesis that matter is a wave. Rather, he considers a classical Hamilton-Jacobi equation for a generic particle and then instead of finding a solution for this equation, he searches for a solution of the associate variational problem, which turns out to be the stationary version of his now-famous equation. While Schrödinger writes that the successful reproduction of the hydrogen spectrum using a wave equation suggests that there is some vibration process in the atom, he however does not push for such an interpretation at this stage. Presumably because he had already in mind some potential problems for this interpretation: in addition to the already mentioned difficulty in interpreting particles as wave packets given that they would quickly spread, the wavefunction (namely the solution of his wave equation) for a system of n classical particles would be a function of $3n$ spatial coordinates and therefore describe a wave in $3n$-dimensional space that could not be identified with ordinary physical space.[13]

In his second paper, "Quantisation as a Proper Value Problem (Part II)," published about a month after the first, Schrödinger provides a more intuitive presentation in terms of the formal analogy (discovered by Hamilton) between geometric optics and classical particle mechanics: both optics and mechanics obey the same variational principle for the same type of characteristic function. This characteristic function had to be minimized, and Fermat's principle of the shortest time and the mechanical principle of least action are just particular cases of Hamilton's more general principle. Schrödinger

---

[13]See L. Wessells (1979) *op. cit.*

uses this analogy to suggest that one *needs* wave mechanics
as a theory of matter: when geometric optics fail, we need
a wave theory in which rays need not to be assumed; similarly we need wave theory when classical mechanics fails and
the notion of path loses its meaning. Then he proceeds to
choose his equation as the simplest equation a wave would
obey. This is the same that he obtained in the first paper,
but now it is coming from very different considerations, as a
product of a new hypothesis about the nature of matter. Here
he mentions again the possibility of interpreting the particle
as a wave packet 'so long as we can neglect any spreading,'
however granting that he has no proof that this approximation is generally valid (he will work more on this in the third
paper of this collection, as we will see below). He reinforces
his realist motivation by emphasizing that in this new picture
one can imagine what is going on inside an atom, and therefore one can understand it. Having arrived to his wave equation through the Hamiltonian analogy, in this second paper
Schrödinger applies it to other cases, such as the Planck oscillator and different varieties of rotators, with success. However he anticipates that a more complete analysis would need
the development of a perturbation theory analog to the one
used in classical mechanics, which he will indeed develop in
the third part of his "Quantisation as a Problem of Proper
Values" series, the fifth paper of the present collection. These
results opened the door to the explanation of the atomic spectra of diatomic molecules.[14]

The third paper, "The Continuous Transition from Micro-
to Macro-Mechanics," is where Schrödinger really works to
make his mathematical model into a physical one. He explores
the possibility of interpreting what we have called up to now
'particles' as localized wave packets. Schrödinger shows that

---

[14]E. Fues (1926) "Zur Intensität der Bandenlinien und des
Affinitätsspektrums zweiatomiger Moleküle," *Annalen der Physik* 81:
281-313. Fues was Schrödinger's assistant in Zürich.

in one dimension a group of suitable waves, intended here as the true microscopic description, can superimpose to form a well-defined and localized 'burst' that does not spread out, so that we could identify it with a 'macroscopic' particle with a definite trajectory. However, Schrödinger admits that the calculations have not been done yet for more realistic cases, for which the situation might change.[15]

The issue taken up in the fourth paper, "On the Relation between the Quantum Mechanics of Heisenberg, Born and Jordan, and that of Schrödinger," is different, but provides another piece of the puzzle that Schrödinger wanted to solve. Before Schrödinger's wave mechanics, various attempts have been proposed to account for experimental data such as the atomic spectra. In 1925, three papers emerged from the collaboration of Heisenberg, Born and Jordan[16] (and the idea was immediately taken up by Dirac[17]) that allowed to describe the experimental results in a general way in terms of infinite matrices, the so-called matrix mechanics. Both matrix mechanics and wave mechanics can account for the

---

[15]Indeed, Lorentz shows that for the hydrogen atom this is not true: letter to Schrödinger, June 19, 1926, in: K. Przibram (ed.), *Letters on wave mechanics*, Martin Klein, trans. Philosophical Library, NY. p. 70. Heisenberg also will make the same point in his uncertainty principle paper. W. Heisenberg (1927): "Ueber den anschaulichen Inhalt der quantentheoretischen Kinematik und Mechanik," *Zeitschrift für Physik*, 43: 172–198. English translation in Wheeler, J.A. and W.H. Zurek (eds), *Quantum Theory and Measurement*, Princeton, NJ: Princeton University Press. 62–84 (1983).

[16]W. Heisenberg (1925): "Über quantentheoretische Umdeutung kinematischer und mechanischer Beziehungen," *Zeitschrift für Physik* 33: 879-893, 1925; M. Born and P. Jordan (1925): "Zur Quantenmechanik." *Zeitschrift für Physik* 34: 858-888; M. Born, W. Heisenberg, and P. Jordan (1926): "Zur Quantenmechanik II." *Zeitschrift für Physik* 35: 557-615.

[17]P. A. M. Dirac (1925): "The Fundamental Equations of Quantum Mechanics," *Proceedings of the Royal Society* 109 (752): 642-653; P. A. M. Dirac (1926):"On the Theory of Quantum Mechanics," *Proceedings of the Royal Society* 112 (762): 661-677.

experimental outcomes, but they are very different from one another on many levels. In his paper Schrödinger points out some differences (continuum vs. discrete picture, algebraic vs. differential equations), but quite strangely he is not explicit here in pointing out what in other correspondence he identified as the difference that mattered to him the most, namely that his wave mechanics is the only one which seems amenable to a straightforward realist interpretation. In contrast with matrix mechanics, his view has the virtue of being visualizable (*anschaulichkeit*): one can see, or imagine, what happens into the atom. The aim of Schrödinger in this paper is to show that these two theories are equivalent reformulation of the same reality. He does that by showing that for each operator in wave mechanics there is one corresponding infinite matrix.[18] Proving this equivalence is important to him because, all other things being equal, the visualizability of his wave mechanics could be taken as a reason to prefer it over matrix mechanics. Indeed, this is the preference expressed by Lorentz after he read the first two of Schrödinger's papers,[19] and this appreciation led to Schrödinger's invitation to the 1927 Solvay conference he was organizing.[20] In addition of excitement, Lorentz however also expressed doubts about wave mechanics, in particular he objected that a wave in configuration space could be thought as a physical wave. Schrödinger was well aware of that, as it is shown by the fact

---

[18] A similar argument was also presented in C. Eckart (1926): "Operator Calculus and the Solution of the Equation of Quantum Dynamics," *Physical Review* 28: 711-726. Another proof of the equivalency has been developed also by Pauli in a letter to Jordan (April 12, 1926), as reported by J. Mehra (1988): "Erwin Schrödinger and the Rise of Wave Mechanics. III. Early Response and Applications," *Foundations of Physics* 18 (2): 107-184.

[19] K. Przibram (1967) *op. cit.*

[20] G. Bacciagaluppi and A. Valentini (2009): *Quantum Theory at the Crossroads: Reconsidering the 1927 Solvay Conference*, Cambridge University Press.

that he acknowledges this problem again at the end of this
paper.

Schrödinger nonetheless is hopeful of solving these problems, and continues in his quest for an intuitive understanding of the quantum world in the fifth paper of this collection,
"Quantisation as a Problem of Proper Values (Part III)." In
this article Schrödinger develops the theory of perturbation
mentioned in Part III to apply the theory beyond directly
solvable problems. He discusses the Stark effect, the shift-
ing and splitting of Balmer spectral lines (the most intense
lines of the Hydrogen spectrum) due to the presence of an
external electric field. He does not discusses the Zeemean effect, an analogous splitting due to a magnetic field, because
as mentioned in the first paper magnetic effects, connected to
relativistic effects, and they cannot yet be accounted for.[21]

The next paper, "Quantisation as a Problem of Proper
Values (Part IV)," marks another stepping stone, as it contains the Schrödinger wave equation in its most general form,
a proposal for a relativistic wave equation, and a new proposal
on how to think of the wavefunction which would solve the
problem of being realist about a wave in configuration space.
The stated aim of the paper is to generalize the equation proposed in the previous papers, which holds only for a fixed
value of energy and it is time-dependent only through the
phase. As such, Schrödinger claims, it is not really a vibration equation, so he embarks in eliminating the energy from
his previous equation. By doing that he first finds a fourth
order equation. Aside from the complexity of it, he observes
that, given that for time independent potentials the equation
is equivalent to the product of two second-order equations,
one could instead consider a fundamnetal wave equation of the
second order instead. This is the Schrödinger equation which

---

[21]This effect will be later accounted for by Fock, who develops a
relativistic wave equation: V. Fock (1926): "Zur Schrödingerschen
Wellenmechanik," *Zeitschrift für Physik* 28: 242- 250.

was about to become famous. Nonetheless, he notices, the price to pay for this is that the the solution of this equation has to be complex. After generalizing perturbation theory to this case, in paragraph 6 Schrödinger constructs a relativistic equation for the single electron by substituting quantities with operators. He remarks that more justification needs to be provided for this equation, not only because it needs to be corrected for spin, but also because he did not want to rely on such a formal analogy, justified only in virtue of the fact that it would reproduce the nonrelativistic equation with the right Hamiltonian. He therefore moves to the physical significance of the solution of his equation. To respond to the objection that a field in configuration space cannot be considered physically vibrating and convinced that the wavefunction had some electromagnetic meaning, he proposes that the square of the wave function is to be interpreted as a charge density, reinforcing the idea that what we call 'particles' are actually localized wave packets.

The last three papers were written in 1927 and constitute further development of Schrödinger's project. The first of these papers, "On the Compton Effect," provides a partial answer to a criticism that he had received by many, including Bohr and Heisenberg during his visit to Copenhagen in October 1926, namely that wave mechanics, being a theory of continuum, cannot account for the distinctive feature of quantum theory, namely that phenomena have a discrete character (the so-called 'quantum jumps'). The idea was that the photoelectric effect (when a material emits charged particles after absorbing radiation) and the Compton effect (the frequency shift of electromagnetic radiation after interaction with matter) could only be explained in terms of light quanta.[22] In re-

---

[22] A. Einstein (1905): "Über einen die Erzeugung und Verwandlung des Lichtes betreffenden heuristischen Gesichttspunkt," *Annalen der Physik* 17: 132-148; A. Compton (1923): "A Quantum Theory of the Scattering of X-rays by Light Elements," *Physical Review* 21 (5): 483-502.

sponse, Schrödinger provides an account of the phenomenon in terms of wave mechanics. He follows the work of Gordon, which developed a relativistic wave equation and already described the Compton effect in wave-mechanical terms,[23] but in a less technical way. Schrödinger argues that the incoming radiation is diffracted on a standing 'charge density' wave, namely the electron, just as light is diffracted on a standing acoustic wave. Around the same time Wentzel[24] proposed an analysis of the photoelectric effect within wave mechanics, and later Mott developed a general method to account for any kind of discontinuous phenomena using wave mechanics, clearing out the road to Schrödinger regarding the objection that one needs a discrete ontology to account for quantum phenomena.[25]

Also the last two papers in the collection use the relativistic extension of the wave theory. In the second to last, "The Energy-Momentum Theorem for Material Waves," Schrödinger wants to see whether one can blend together quantum theory and classical electrodynamics using the Hamilton principle that was used by some to derive the relativistic wave equation,[26] and emphasizes many of the difficulties on en-

---

[23]W. Gordon (1926): "Der Comptoneffekt nach der Schrödingerschen Theorie," *Zeitschrift für Physik* 40 (1–2): 117-133. The relativistic form of the wave equation was developed by several people independently: O. Klein (1926): "Quantentheorie und Fünfdimensionale Relativitätstheorie," *Zeitschrift für Physik* 37 (12): 895-906; V. Fock, op. cit.; J. Kudar (1926): "Zur vierdimensionalen Formulierung der undulatorischen Mechanik," *Annalen der Physik*, 8 : 632–636; L. de Broglie (1926): "Remarques sur la Nouvelle Mécanique Ondulatoire," *Comptes rendus de l'Académie des Sciences* 183: 272-3; T. de Donder, H. van der Dungen (1926): "La Quantification Déduite de la Gravifique Einsteinienne," *Comptes rendus de l'Académie des Sciences* 183: 22–24. See also Pauli, in the previously cited letter to Jordan.

[24]G. Wentzel (1926): "Zur Theorie des Photoelektrischen Effekts," *Zeitschrift für Physik* 40:574-589.

[25]N.F. Mott (1929): "The Wave Mechanics of α-ray Tracks," *Proceedings of the Royal Society* A 126: 79-84.

[26]O. Klein (1926), *op.cit.*; T. de Donder, H. van der Dungen (1926),

counter in doing such a project (the theory gives the wrong results when applied to the hydrogen atom). In the final paper of this volume, "The Exchange of Energy according to Wave Mechanics," Schrödinger wants again to defend his view against the criticism that his theory cannot account for 'quantum jumps.' Schrödinger thus shows that two atoms in resonance exchange energy *as if* they were exchanging discrete quanta. He proposes that we should re-understand quanta in terms of wave frequency, and that the notion of resonance is key to understand quantum phenomena. Also, he provides a wave-mechanical demonstration of Planck's radiation law without postulating the existence of light quanta. He acknowledges that one could interpret these results in terms of probablity field rather than material fields, as Born had just done,[27] but he rejects it somewhat cryptically on grounds that it is not sufficiently warranted. It does not seem that he disliked determinism: early in his carreer he supported such a view, as shown by his work on the BKS (Bohr, Kramers, Slater) theory of radiation, where momentum and energy were conserved only statistically.[28] Rather, Schrödinger's problem with Born's statistical interpretation of the wavefunction seems to be that one does not gain much from it, as a matter of explanation. He does not elaborate on what the problem is supposed to be in this paper but in a letter to Planck he writes: "What seems most questionable to me in Born's probability interpretation is that [...] the probabilities of events that a naïve interpretation would consider to be independent do not simply multiply when combined, but instead the 'probability amplitudes interfere' in a completely mysterious way (namely, just like my wave amplitudes, of course)."[29] In this

op. cit.

[27]M. Born (1926): "Zur Quantenmechanik der Stoßvorgänge," *Zeitschrift für Physik* 37 (12): 863–867.

[28]N. Bohr, H.A. Kramers, J.C. Slater (1924), *op. cit.*

[29]E. Schrödinger, Letter to Planck, July 4, 1927, in: K. Przibrarn

way Schrödinger anticipates the strongest objection to the epistemic views of the wavefunction: how can one explain the real phenomena of diffraction and interference if the wavefunction represents our state of knowledge of reality, rather than reality itself?

This paper concludes the collection, both practically and pragmatically, as it marks the end of Schrödinger realist project. In 1927 Heisenberg wrote his celebrated uncertainty principle paper, in which his main goal was to defend his matrix mechanics from Schrödinger's charge of lack of visualizability. Heisenberg therefore endorses a partcile picture of reality, and argues that by abandoing the idea of trajectory then everything becomes comprehensible and intutitive. In this way he *de facto* nullifies the advantage of wave mechanics provided by Schrödinger's proof of equivalency between matrix and wave mechanics, as they are both visualizable. Incidentally, Heisenberg's paper made Bohr very upset, becasue he wanted to keep the wave picture together with the partcile picture.[30] In fact Bohr disliked the light quanta and used wave concepts in his theories: he developed the BKS theory to dispense of light quanta, and his atomic theory in terms of stationary states was more compatible with Schrödinger's wave theory than the partcile theory favored by Heisenberg. However, Bohr insisted in reading Schrödinger's picture instrumentally, and begun to develop what was to become his theory of complementarity: wave and particles are concepts which complement each other, and show their face one at a time. As a result, he strongly pressed Heisenberg to change his mind and endorse his view, so that the 'Copehagen-Göttingen school' formed a united front against Schroedinger's wave picture. The establishment of the Copehagen dogma culminated in 1928, when Bohr delivered the now famous Como lecture, which

(1967), *op. cit.* p.20.

[30]M. Beller (1996): "The Conceptual and the Anecdotal History of Quantum Mechanics," *Foundations of Physics* 26 (4): 545-557.

provides the basis of the complementarity view. On another front Dirac[31] simply side-stepped the question of the relative physical realism by introducing wave and matrix mechanics as calculational aids, and emphasizing his own symbolic approach for its ability to express the physical laws in a neat and concise way. Eventually Schrödinger changed research topics to study a way of unifying quantum theory and relativity using fields, along the line of research developed by Einstein, with which he created a long and fruitful collaboration. He returned to quantum theory sporadically in print, with the notable exception of 1935, when after the publication of the Einstein-Podolsky-Rosen paper[32] Schrödinger wrote a paper entitled "The Present Situation in Quantum Mechanics"[33] in which, among other things, he presented his now famous cat which is impossibly dead and alive as a *reductio ad absurdum* for the theory.

---

[31]P. A. M. Dirac (1925), op. cit.

[32]A. Einstein, N. Podolsky, N. Rosen (1935): "Can Quantum-Mechanical Description of Physical Reality be Considered Complete?" *Physical Review* 47 (10): 777–780.

[33]E. Schrödinger (1935): "Die Gegenwärtige Situation in der Quantenmechanik," *Naturwissenscahften* 32 (48): 807-812. Translation by J. D. Trimmer (1980): "The Present Situation in Quantum Mechanics: A Translation of Schrödinger's "Cat Paradox" Paper," *Proceedings of the American Philosophical Society* 124 (5): 323-338.

xx

Preface to the First (German) Edition

Referring to these six papers (the present reprint of which is solely due to the great demand for separate copies), a young lady friend recently remarked to the author: "When you began this work you had no idea that anything so clever would come out of it, had you?" This remark, with which I wholeheartedly agreed (with due qualification of the flattering adjective), may serve to call attention to the fact that the papers now combined in one volume were originally written *one by one* at different times. The results of the later sections were largely unknown to the writer of the earlier ones. Consequently, the material has unfortunately not always been set forth in as orderly and systematic a way as might be desired, and further, the papers exhibit a gradual development of ideas which (owing to the nature of the process of reproduction) could not be allowed for by any alteration or elaboration of the earlier sections. The *Abstract* which is prefixed to the text may help to make up for these deficiencies.

The fact that the papers have been reprinted without alteration in no way implies that I claim to have succeeded in establishing a theory which, though capable of (and indeed requiring) extension, is firmly based as regards its physical foundations and henceforth admits of no alteration in its fundamental ideas. On the contrary, this comparatively cheap method of issue seemed advisable on account of the impossibility at the present stage of giving a fresh exposition which would be really satisfactory or conclusive.

E. Schrödinger

Zürich, *November* 1926

THE ENGLISH PUBLISHER'S NOTE

This translation has been prepared from the second edition of the author's *Abhandlungen zur Wellenmechanik*, published by Johann Ambrosius Barth, 1928. These papers include practically all that Professor Schrödinger has written on Wave Mechanics.

The translation has been made by J. F. Shearer, M.A., B.Sc., of the Department of Natural Philosophy in the University of Glasgow, and W. M. Deans, B.A., B.Sc., late of Newnham College, Cambridge.

The translators have tried to follow the original as closely as the English idiom would permit. The English version has been read by Professor Schrödinger. Throughout the book *Eigenfunklion* has been translated *proper function*, and *Eigenwert*, *proper value*. The phrase *eine stückweise stetige Function* has been translated *a sectionally continuous function*. These equivalents were decided upon after consultation with the author and with several English mathematicians of eminence.

ABSTRACT

The Hamiltonian analogy of mechanics to optics is an analogy to *geometrical* optics, since to the path of the representative point in configuration space there corresponds on the optical side the *light ray*, which is only rigorously defined in terms of geometrical optics. The undulatory elaboration of the optical picture leads to the surrender of the idea of the *path of the system*, as soon as the dimensions of the path are *not* great in comparison with the wave-length. Only when they are so does the idea of the path remain, and with it classical mechanics as an approximation; whereas for "micromechanical" motions the fundamental equations of mechanics are just as useless as geometrical optics is for the treatment of diffraction problems. In analogy with the latter case, a *wave equation* in configuration space must replace the fundamental equations of mechanics, in the first instance, this equation is stated for purely periodic vibrations sinusoidal with respect to time; it may also be derived from a "Hamiltonian variation principle." It contains a "proper value parameter" E, which corresponds to the mechanical energy in macroscopic problems, and which for a single time-sinusoidal vibration is equal to the frequency multiplied by Planck's quantum of action h. In general the wave or vibration equation possesses no solutions, which together with their derivatives are one-valued, finite, and continuous throughout configuration space, *except* for certain special values of E, the *proper values*. These values

1

form the "proper value spectrum" which frequently includes continuous parts (the "band spectrum," not expressly considered in most formulae) as well as discrete points (the "line spectrum"). The proper values either turn out to be identical with the "energy levels" ($=$ spectroscopic "term"-value multiplied by h) of the quantum theory as hitherto developed, or differ from them in a manner which is confirmed by experience. (Unperturbed Keplerian motion; harmonic oscillator; rigid rotator; non-rigid rotator; Stark effect.) Deviations of the kind mentioned are, e.g., the appearance of non-integral quantum numbers (viz. the halves of odd numbers) in the case of the oscillator and rotator, and further, the non-appearance of the "surplus" levels (viz. those with vanishing azimuthal or equatorial quantum number) in the Kepler problem. Even in these matters the agreement with Heisenberg's quantum mechanics is complete: this can be proved in general. For the calculation of the proper values and the corresponding solutions of the vibration equation ("proper functions") in more complicated cases, there is developed a *theory of perturbations*, which enables a more difficult problem to be reduced by quadratures alone to a "neighbouring" but simpler one. To "degeneracy" corresponds the appearance of *multiple* proper values. Especially important physically is the case where, as, e.g., in the Zeeman and Stark effects, a multiple proper value is split up by the addition of perturbing forces.

Up till now the function ψ has merely been defined in a purely formal way as obeying the above-mentioned wave equation, serving as its object, so to speak. It is necessary to ascribe to ψ a physical, namely an electromagnetic, meaning, in order to make the fact that a small mechanical system can emit *electromagnetic* waves of a frequency equal to a term-difference (difference of two proper values divided by h) intelligible at all, and further, in order to obtain a theoretical statement for the intensity and polarisation of these electromagnetic waves. This meaning, for the general case of a system with an arbitrary number of degrees of freedom, is not clearly worked out until the end of the sixth paper. A definite ψ-distribution in configuration space is interpreted as a continuous distribution of electricity (and of electric current density) in actual space. If from this distribution of electricity we calculate the component of the electric moment of the whole system in any direction in the usual way, it appears as the sum of single terms, each of which is associated with a couple of proper vibrations, and vibrates in a purely sinusoidal manner with respect to the time with a frequency equal to. the difference of the allied proper frequencies. If the wave-length of the *electromagnetic* waves, associated with this difference frequency, is large compared with the dimensions of the region to which the whole distribution of electricity is practically confined, then, according to the rules of ordinary electrodynamics, the amplitude of the partial moment in question (or, more accurately, the square of this amplitude multiplied by the fourth power of the frequency) is a measure of the intensity of the light radiated *with* this frequency, and with *this* direction of polarisation. The electrodynamic hypothesis concerning ψ, and the related purely classical calculation of the radiation, are verified by experience, in so far as they furnish the customary *selection and polarisation rules* for the oscillator, rotator and the hydrogen atom. Further, they also furnish satisfactory intensity relations for the fine structure of the Balmer lines in an electric field. If only *one* proper vibration or only proper vibrations of *one* proper frequency are excited, then the electrical distribution becomes *static*, yet *stationary currents* may possibly be superimposed. In this manner the *stability* of the normal state and its *lack of radiation* are explained.

The amplitudes of the partial moments are closely connected with those quantities ("matrix elements"), which determine the radiation, according to the formal theory of Heisenberg, Born, and Jordan. There can be demonstrated a far-

4

reaching formal identity of the two theories, according to which not only do the calculated emission frequencies and selection and polarisation rules agree, but also the above-mentioned successful results of the intensity calculations are to be credited as much to the matrix theory as to the present one.

Everything up till now has referred in the first instance only to *conservative* systems, although some parts have reached their final formulation only in the sixth paper in connection with the treatment of non-conservative systems. For the latter, the wave equation used hitherto must be generalised into a *true* wave equation, which contains the time explicitly, and is valid not merely for vibrations purely sinusoidal with respect to time (with a frequency which appears in the equation as a proper value parameter), but for any arbitrary dependence on the time. From the wave equation generalised in this way, the interaction of the system with an incident light wave can be deduced, and hence a rational dispersion formula; in all this the electrodynamic hypothesis about ψ is retained. The generalisation for an arbitrary disturbance is indicated. Further, from the generalised wave equation an interesting conservation theorem for the "weight function" $\psi\bar{\psi}$ can be obtained, which demonstrates the complete justification of the electrodynamic hypothesis frequently mentioned above, and which makes possible the deduction of the expressions for the components of the electric current density, in terms of the ψ-distribution.

Even the systems treated in the first five papers cannot be conservative in the literal sense of the word, inasmuch as they radiate energy; this must be accompanied by a change in the system. Thus there still seems to be something lacking in the wave law for the function, – corresponding to the "reaction of radiation" of the classical electron theory, which may result in a dying away of the higher vibrations in favour of the lower ones. This necessary complement is still *missing*.

The form of the theory discussed so far corresponds to classical (i.e. non-relativistic) mechanics, and does not take magnetic fields into consideration. Therefore, neither the wave equation nor the components of the four-current are invariant for the Lorentz transformation. For the *one*-electron problem an immediate relativistic- magnetic generalisation is readily suggested (the Lorentz-invariant expressions for the components of the four-current are not given in the text, but they can be got[34] from the "equation of continuity," which is to be formed in a way quite analogous to that in the non-relativistic case). Though this generalisation yields *formally* reasonable expressions for the wave lengths, polarisations, intensities, and selection in the natural fine structure and in the Zeeman pattern of the hydrogen atom, yet the actual diagram turns out quite wrong, for the reason that "half integers" appear as azimuthal quantum numbers in the Sommerfeld fine structure formula (V. Fock carried out the calculations quite independently in Leningrad, before my last paper was sent in, and also succeeded in deriving the relativistic equation from a variation principle. *Zeitschrift für Physils*, 38, p. 242, 1926). A correction is therefore necessary; all that can be said about it at present is that it must have the same significance for wave mechanics as the "spinning electron" of Uhlenbeck and Goudsmit has for the older quantum theory dealing with electronic orbits; with this difference, however, that in the latter, together with the introduction of the "spinning electron," the half-integral form of the azimuthal quantum number must be postulated *ad hoc*, in order to avoid serious conflict with experiment even in the case of hydrogen; while wave mechanics (and also Heisenberg's quantum mechanics) *necessarily* yields halves of odd integers (German: Halbzahligkeit), and thus gives a hint, from the very beginning, of that further extension, which under the regime of the older theory was only

---

[34]Cf. also a paper by W. Cordon on the Compton Effect, *Ztschr. f. Phys.* 40, p. 117, 1926.

shown to be necessary by more complicated phenomena, such as the Paschen-Back effect in hydrogen, anomalous Zeeman effects, structures of multiplets, the laws of Röntgen doublets and the analogy between them and the alkali doublets.

Addition in the second (German) edition: the first and second of the three new papers now added, namely, "The Compton Effect" and "The Energy-Momentum Theorem for Material Waves," are contributions to the four-dimensional relativistic form of wave mechanics discussed in the above paragraph. In connection with the first of these papers I should like above all to remark that, as Herr Ehrenfest has pointed out to me, the figure (p. 202) is incorrect: the pair of wave trains represented in the right half of the figure should coincide *completely* with the pair on the left, in respect of wave length and the orientation of their planes as well as in breadth of interference fringes (the broken lines). The second paper, that on "The Energy-Momentum Theorem," throws a strong light on the difficulties which a merely *four*-dimensional theory of ψ-waves comes up against, despite the formally beautiful possibilities of development which present themselves here. In the last paper, on "The Exchange of Energy according to Wave Mechanics," the *many*-dimensional, non- relativistic form is again used. This paper is a first attempt to find out whether, with reference to Heisenberg's important discovery of the "quantum mechanics resonance phenomenon," it should not be possible to regard those very phenomena which seem to be decisive evidence for the existence of discrete energy levels, without this hypothesis, *merely* as resonance phenomena.

QUANTISATION AS A PROBLEM OF PROPER VALUES (PART I)

(*Annalen der Physik* (4), vol. 79, 1926)

§ 1. In this paper I wish to consider, first, the simple case of the hydrogen atom (non-relativistic and unperturbed), and show that the customary quantum conditions can be replaced by another postulate, in which the notion of "whole numbers," merely as such, is not intro- duced. Rather when integralness does appear, it arises in the same natural way as it does in the case of the *node-numbers* of a vibrating string. The new conception is capable of generalisation, and strikes, I believe, very deeply at the true nature of the quantum rules.

The usual form of the latter is connected with the Hamilton-Jacobi differential equation,

$$H\left(q, \frac{\partial S}{\partial q}\right) = E. \tag{1}$$

A solution of this equation is sought such as can be represented as the *sum* of functions, each being a function of one only of the independent variables q.

Here we now put for S a new unknown ip such that it will appear as a 'product of related functions of the single co-ordinates, i.e. we put

$$S = K \log \psi. \tag{2}$$

The constant K must be introduced from considerations

7

of dimensions; it has those of *action.* Hence we get

$$H\left(q, \; \frac{K}{\psi}\frac{\partial\psi}{\partial q}\right) = E. \tag{1'}$$

Now we do *not* look for a solution of equation (1′), but proceed as follows. If we neglect the relativistic variation of mass, equation (1′) can always be transformed so as to become a quadratic form (of ψ and its first derivatives) equated to zero. (For the *one*-electron problem this holds even when mass-variation is not neglected.) We now seek a function ψ, such that for any arbitrary variation of it the integral of the said quadratic form, taken over the whole coordinate space,[1] is stationary, ψ being everywhere real, single-valued, finite, and continuously differentiable up to the second order. *The quantum conditions are replaced by this variation problem.*

First, we will take for H the Hamilton function for Keplerian motion, and show that ψ can be so chosen for *all positive,* but only for a *discrete set of negative values of E.* That is, the above variation problem has a discrete and a continuous spectrum of proper values.

The discrete spectrum corresponds to the Balmer terms and the continuous to the energies of the hyperbolic orbits. For numerical agreement K must have the value $h/2\pi$.

The choice of .co-ordinates in the formation of the variational equa- tions being arbitrary, let us take rectangular Cartesians. Then (1′) becomes in our case

$$\left(\frac{\partial\psi}{\partial x}\right)^2 + \left(\frac{\partial\psi}{\partial y}\right)^2 + \left(\frac{\partial\psi}{\partial z}\right)^2 - \frac{2m}{K^2}\left(E + \frac{e^2}{r}\right)\psi^2 = 0; \tag{1''}$$

$e = $ charge, $m = $ mass of an electron, $r^2 = x^2 + y^2 + z^2.$

---

[1] I am aware this formulation is not entirely unambiguous.

Our variation problem then reads

$$\delta J = \delta \iiint dx\, dy\, dz \times$$

$$\times \left[\left(\frac{\partial \psi}{\partial x}\right)^2 + \left(\frac{\partial \psi}{\partial y}\right)^2 + \left(\frac{\partial \psi}{\partial z}\right)^2 - \frac{2m}{K^2}\left(E + \frac{e^2}{r}\right)\psi^2 \right] = 0,$$

(3)

the integral being taken over all space. From this we find in the usual way

$$\frac{1}{2}\delta J = \int df\, \delta\psi \frac{\partial \psi}{\partial n} -$$

$$- \iiint dx\, dy\, dz\, \delta\psi \left[\nabla^2 \psi + \frac{2m}{K^2}\left(E + \frac{e^2}{r}\right)\psi \right] = 0.$$

(4)

Therefore we must have, firstly,

$$\nabla^2 \psi + \frac{2m}{K^2}\left(E + \frac{e^2}{r}\right)\psi = 0,$$

(5)

and secondly,

$$\int df\, \delta\psi \frac{\partial \psi}{\partial n} = 0.$$

(6)

df is an element of the infinite closed surface over which the integral is taken.

(It will turn out later that this last condition requires us to supplement our problem by a postulate as to the behaviour of $\delta\psi$ at infinity, in order to ensure the existence of the above-mentioned continuous spectrum of proper values. See later.)

The solution of (5) can be effected, *for example*, in polar coordinates, r, θ, ϕ, if ψ be written as the *product* of three functions, each only of r, of θ, or of ϕ. The method is sufficiently well known. The function of the angles turns out to be a *surface harmonic*, and if that of r be called χ we get easily the differential equation,

$$\frac{d^2\chi}{dr^2} + \frac{2}{r}\frac{d\chi}{dr} + \left(\frac{2mE}{K^2} + \frac{2me^2}{K^2 r} - \frac{n(n+1)}{r^2} \right)\chi = 0. \quad (7)$$

$$n = 0, 1, 2, 3 \ldots$$

The limitation of n to integral values is *necessary* so that the surface harmonic may be *single-valued*. We require solutions of (7) that will remain finite for all non-negative real values of r. Now[2] equation (7) has *two* singularities in the complex $r-$plane, at $r = 0$ and $r = \infty$, of which the second is an "indefinite point" (essential singularity) of all integrals, but the first on the contrary is not (for any integral). These two singularities form exactly the *bounding points of our real interval*. In such a case it is known now that the postulation of the *finiteness* of χ the bounding points is equivalent to a *boundary condition*. The equation has *in general* no integral which remains finite at *both* end points; such an integral exists only for certain special values of the constants in the equation. It is now a question of defining these special values. This is the *jumping-off* point of the whole investigation.[3]

Let us examine first the singularity at $r = 0$. The so-called *indicial* equation which defines the behaviour of the integral at this point, is

$$\rho(\rho - 1) + 2\rho - n(n + 1) = 0, \tag{8}$$

with roots

$$\rho_1 = n, \qquad \rho_2 = -(n + 1). \tag{8'}$$

The two canonical integrals at this point have therefore the exponents n and $-(n + 1)$. Since n is not negative, only the first of these is of use to us. Since it belongs to the greater exponent, it can be represented by an ordinary power series, which begins with r^n. (The other integral, which does not interest us, can contain a logarithm, since the difference between the indices is an integer.) The next singularity is at

---

[2]For guidance in the treatment of (7) I owe thanks to Hermann Weyl.
[3]For unproved propositions in what follows, see L. Schlesinger's *Differential Equations* (Collection Schubert, No. 13, Göschen, 1900, especially chapters 3 and 5).

infinity, so the above power series is always convergent and represents a *transcendental integral function*. We therefore have established that:

The required solution is (except for a constant factor) a single-valued definite transcendental integral function, which at r = 0 belongs to the exponent n.

We must now investigate the behaviour of this function at infinity on the positive real axis. To that end we simplify equation (7) by the substitution

$$\chi = r^{\alpha} U \tag{9}$$

where a is so chosen that the term with $1/r^2$ drops out. It is easy to verify that then α must have one of the two values n, $-(n+1)$. Equation (7) then takes the form,

$$\frac{d^2 U}{dr^2} + \frac{2(\alpha+1)}{r}\frac{dU}{dr} + \frac{2m}{K^2}\left(E + \frac{e^2}{r}\right)U = 0. \tag{7'}$$

Its integrals belong at $r = 0$ to the exponents 0 and $-2\alpha - 1$. For the α-value, $\alpha = n$, the *first* of these integrals, and for the second α-value, $\alpha = -(n+1)$, the *second* of these integrals is an integral function and leads, according to (9), to the desired solution, which is single-valued. We therefore lose nothing if we confine ourselves to one of the two α-values. Take, then,

$$\alpha = n. \tag{10}$$

Our solution U then, at $r = 0$, belongs to the exponent 0. Equation (7') is called Laplace's equation. The general type is

$$U'' + \left(\delta_0 + \frac{\delta_1}{r}\right)U' + \left(\epsilon_0 + \frac{\epsilon_1}{r}\right)U = 0. \tag{7''}$$

Here the constants have the values

$$\delta_0 = 0, \quad \delta_1 = 2(\alpha+1), \quad \epsilon_0 = \frac{2mE}{K^2}, \quad \epsilon_1 = \frac{2me^2}{K^2}. \tag{11}$$

This type of equation is comparatively simple to handle for this reason: The so-called Laplace's transformation, which in general leads *again* to an equation of the *second* order, *here* gives one of the *first*. This allows the solutions of (7″) to be represented by complex integrals. The result[4] only is given here. The integral

$$U = \int_L e^{zr}(z - c_1)^{\alpha_1 - 1}(z - c_2)^{\alpha_2 - 1}dz \qquad (12)$$

is a solution of (7″) for a path of integration L, for which

$$\int_L \frac{d}{dz}\left[e^{zr}(z - c_1)^{\alpha_1}(z - c_2)^{\alpha_2}\right]dz \qquad (13)$$

The constants $c_1, c_2 \alpha_1, \alpha_2$ have the following values. c_1 and c_2 are the roots of the quadratic equation

$$z^2 + \delta_0 z + \epsilon_0 = 0, \qquad (14)$$

and

$$\alpha_1 = \frac{\epsilon_1 + \delta_1 c_1}{c_1 - c_2}, \qquad \alpha_2 = \frac{\epsilon_1 + \delta_1 c_2}{c_1 - c_2} \qquad (14')$$

In the case of equation (7′) these become, using (11) and (10),

$$c_1 = +\sqrt{\frac{-2mE}{K^2}}, \qquad c_2 = -\sqrt{\frac{-2mE}{K^2}}; \qquad (14'')$$

$$\alpha_1 = \frac{me^2}{K\sqrt{-2mE}} + n + 1, \qquad \alpha_2 = -\frac{me^2}{K\sqrt{-2mE}} + n + 1.$$

The representation by the integral (12) allows us, not only to survey the asymptotic behaviour of the totality of solutions when r tends to infinity in a definite way, but also to give an account of this behaviour for one *definite* solution, which is always a much more difficult task.

---

[4]Cf. Schlesinger. The theory is due to H. Poincaré and J. Horn.

We shall at first *exclude* the case where α_1 and α_2 are real integers. When this occurs, it occurs for both quantities simultaneously, and when, and only when,

$$\frac{me^2}{K\sqrt[+]{-2mE}} = \text{a real integer.} \tag{15}$$

Therefore we assume that (15) is not fulfilled.

The behaviour of the totality of solutions when r tends to infinity in a definite manner – we think always of r becoming infinite through real positive values – is characterised[5] by the behaviour of the two linearly independent solutions, which we will call U_1 and U_2, and which are obtained by the following *specialisations* of the path of integration L. In *each* case let z come from infinity and return there along the same path, in such a direction that

$$\lim_{z \to \infty} e^{zr} = 0, \tag{16}$$

the real part of zr is to become negative and infinite. In this way condition (13) is satisfied. In the *one* case let z make a circuit once round the point c_1 (solution U_1), and in the *other*, round c_2 (solution U_2).

Now for very large real positive values of r, these two solutions are represented asymptotically (in the sense used by Poincaré) by

$$\begin{cases} U_1 \sim e^{c_1 r} r^{-\alpha_1} (-1)^{\alpha_1} \left(e^{2\pi i \alpha_1} - 1 \right) \Gamma(\alpha_1)(c_1 - c_2)^{\alpha_2 - 1}, \\ U_2 \sim e^{c_2 r} r^{-\alpha_2} (-1)^{\alpha_2} \left(e^{2\pi i \alpha_2} - 1 \right) \Gamma(\alpha_2)(c_2 - c_1)^{\alpha_1 - 1}, \end{cases} \tag{17}$$

in which we are content to take the first term of the asymptotic series of integral negative powers of r.

We have now to distinguish between the two cases.

---

[5] If (15) is satisfied, at least one of the two paths of integration described in the text cannot be used, as it yields a vanishing result.

1. $E > 0$. This guarantees the non-fulfilment of (15), as it makes the left hand a pure imaginary. Further, by (14"), c_1 and c_2 also become pure imaginaries. The exponential functions in (17), since r is real, are therefore periodic functions which remain finite. The values of α_1 and α_2 from (14") show that both U_1 and U_2 tend to zero like r^{-n-1}. *This must therefore be valid for our transcendental integral solution U,* whose behaviour we are investigating, *however it may be linearly compounded from U_1 and U_2.* Further, (9) and (10) show that the function χ, i.e. the transcendental integral solution of the *original* equation (7), always tends to zero like $1/r$, as it arises from U through multiplication by r^n. We can thus state:

The Eulerian differential equation (5) of our variation problem has, for every positive E, solutions, which are everywhere single-valued, finite, and continuous; and which tend to zero with $1/r$ at infinity, under continual oscillations. The surface condition (6) has yet to be discussed.

2. $E < 0$. In this case the possibility (15) is not *eo ipso* excluded, yet we will maintain that exclusion provisionally. Then by (14") and (17), for $r \to \infty$, U_1 grows beyond all limits, but U_2 vanishes exponentially. Our integral function U (and the same is true for χ) will then remain finite if, and only if, V is identical with f/ 2, save perhaps for a numerical factor. *This, however, can never be,* as is proved thus: If a *closed* circuit round *both* points c_1 and c_2 be chosen for the path L, thereby satisfying condition (13) since the circuit is *really closed* on the Riemann surface of the integrand, on account of $\alpha_1 + \alpha_2$ being an integer, then it is easy to show that the integral (12) represents our *integral function U*. (12) can be developed in a series of positive powers of r, which converges, at all events, for r sufficiently small, and since it satisfies equation (7′), it must coincide with the series for U. Therefore U is represented by (12) if L be a closed circuit round both points c_1 and c_2. This closed circuit can be so distorted,

however, as to make it appear additively combined from the two paths, considered above, which belonged to U_1 and U_2; and the factors are non-vanishing, 1 and $e^{2\pi i \alpha_1}$. Therefore U cannot coincide with U_2, but must contain also U_1. Q.E.D.

Our integral function U, which alone of the solutions of $(7')$ is considered for our problem, is therefore not finite for r large, on the above hypothesis. Reserving meanwhile the question of *completeness*, *i.e.* the proving that our treatment allows us to find all the linearly independent solutions of the problem, then we may state:

For negative values of E which do not satisfy condition (15) our variation problem has no solution.

We have now only to investigate that discrete set of negative E-values which satisfy condition (15). α_1 and α_2 are then both integers. The first of the integration paths, which previously gave us the funda- mental values U_1 and U_2, must now undoubtedly be modified so as to give a non-vanishing result. For, since $\alpha_1 - 1$ is certainly positive, the point c_1 is neither a branch point nor a pole of the integrand, but an ordinary zero. The point c_2 can also become regular if $\alpha_2 - 1$ is also not negative. In *every* case, however, two suitable paths are readily found and the integration effected completely in terms of known functions, so that the behaviour of the solutions can be fully investigated.

Let

$$\frac{me^2}{K\sqrt{-2mE}} = l; \qquad l = 1, 2, 3, 4 \dots \qquad (15')$$

Then from $(14'')$ we have

$$\alpha_1 - 1 = l + n, \qquad \alpha_2 - 1 = -l + n. \qquad (14''')$$

Two cases have to be distinguished: $l < n$ and $l > n$.

(a) $l < n$. Then c_2 *and* c_1 lose every singular character, but instead become starting-points or end-points of the path of integration, in order to fulfil condition (13). A third characteristic point here is at infinity (negative and real). Every path between two of these three points yields a solution, and of these three solutions there are two linearly independent, as is easily confirmed if the integrals are calculated out. In particular, the *transcendental integral solution* is given by the path from c_1 to c_2. That *this* integral remains regular at $r = 0$ can be seen at once without calculating it. I emphasize this point, as the actual calculation is apt to obscure it. However, the calculation does show that the integral becomes indefinitely great for positive, infinitely great values of r. One of the *other* two integrals remains *finite* for r large, but it becomes infinite for $r = 0$.

Therefore when $l < n$ we get no solution of the problem.

(b) $l > n$. Then from (14'''), c_1 is a zero and c_2 a pole of the first order at least of the integrand. Two independent integrals are then obtained: one from the path which leads from $z = -\infty$ to the zero, intentionally avoiding the pole; and the other from the *residue* at the pole. The *latter* is the integral function. We will give its calculated value, but multiplied by r'', so that we obtain, according to (9) and (10), the solution x of the original equation (7). (The multiplying constant is arbitrary.) We find

$$\chi = f\left(r \frac{\sqrt{-2mE}}{K}\right);$$

$$f(x) = x^n e^{-x} \sum_{k=0}^{l-n-1} \frac{(-2x)^k}{k!} \binom{l+n}{l-n-1-k}. \tag{18}$$

It is seen that this is a solution that can be utilised, since it remains finite for all real non-negative values of r. In addition, it satisfies the surface condition (6) because of its vanishing exponentially at infinity. Collecting then the results for E negative:

For E negative, our variation problem has solutions if, and only if, E satisfies condition (15). Only values smaller

than l (and there is always at least one such at our disposal)
can be given to the integer n, which denotes the order of the
surface harmonic appearing in the equation. The part of the
solution depending on r is given by (18).

Taking into account the constants in the surface harmonic
(known to be $2n + 1$ in number), it is further found that:

The discovered solution has exactly $2n + 1$ arbitrary con-
stants for any permissible (n, l) combination; and therefore
for a prescribed value of l has l^2 arbitrary constants.

We have thus confirmed the main points of the statements
originally made about the proper-value spectrum of our variation problem, but there are still deficiencies.

Firstly, we require information as to the completeness of
the *collected* system of proper functions indicated above, but
I will not concern myself with that in this paper. From experience of similar cases, it may be supposed that no proper
value has escaped us.

Secondly, it must be remembered that the proper functions, ascertained for E positive, do not solve the variation
problem as originally postulated, because they only tend to
zero at infinity as $1/r$, and therefore $\partial\psi/\partial r$ only tends to
zero on an infinite sphere as $1/r^2$. Hence the surface integral (6) is still of the same order as $\delta\psi$ at infinity. If it is
desired therefore to obtain the continuous spectrum, another
condition must be added to the *problem*, viz. that $\delta\psi$ is to
vanish at infinity, or at least, that it tends to a constant value
independent of the direction of proceeding to infinity; in the
latter case the surface harmonics cause the surface integral to
vanish.

§ 2. Condition (15) yields

$$- E_l = \frac{me^4}{2K^2 l^2}. \tag{19}$$

Therefore the well-known Bohr energy-levels, corresponding
to the Balmer terms, are obtained, if to the constant K, in-

18

troduced into (2) for reasons of dimensions, we give the value

$$K = \frac{h}{2\pi},$$
(20)

from which comes

$$- E_l = \frac{2\pi m e^4}{h^2 l^2}.$$
(19')

Our l is the principal quantum number, $n + 1$ is analogous to the azimuthal quantum number. The splitting up of
this number through a closer definition of the surface harmonic can be compared with the resolution of the azimuthal
quantum into an "equatorial" and a "polar" quantum. These
numbers *here* define the system of node-lines on the sphere.
Also the "radial quantum number" $l - n - 1$ gives exactly the
number of the "node-spheres," for it is easily established that
the function $f(x)$ in (18) has exactly $l - n - 1$ positive real
roots. The positive E-values correspond to the continuum of
the hyperbolic orbits, to which one may ascribe, in a certain
sense, the radial quantum number ∞. The fact corresponding
to this is the proceeding to infinity, under *continual* oscillations, of the functions in question.

It is interesting to note that the range, inside which the
functions of (18) differ sensibly from zero, and outside which
their oscillations die away, is of the *general order of magnitude*
of the major axis of the ellipse in each case. The factor,
multiplied by which the radius vector enters as the argument
of the constant-free function f, is – naturally – the reciprocal
of a length, and this length is

$$\frac{K}{\sqrt{-2mE}} = \frac{K^2 l}{me^2} = \frac{h^2 l}{4\pi^2 me^2} = \frac{a_l}{l},$$
(21)

where $a_l =$ the semi-axis of the lth elliptic orbit. (The equations follow from (19) plus the known relation $E_l = \frac{-e^2}{2a_l}$).

The quantity (21) gives the order of magnitude of the range of the roots when l and n are small; for then it may be assumed that the roots of $f(x)$ are of the order of unity. That is naturally no longer the case if the coefficients of the polynomial are large numbers. At present I will not enter into a more exact evaluation of the roots, though I believe it would confirm the above assertion pretty thoroughly.

§ 3. It is, of course, strongly suggested that we should try to connect the function i/j with some vibration process in the atom, which would more nearly approach reality than the electronic orbits, the real existence of which is being very much questioned to-day. I originally intended to found the new quantum conditions in this more intuitive manner, but finally gave them the above neutral mathematical form, because it brings more clearly to light what is really essential. The essential thing seems to me to be, that the postulation of "whole numbers" no longer enters into the quantum rules mysteriously, but that we have traced the matter a step further back, and found the "integralness" to have its origin in the finiteness and single-valuedness of a certain space function.

I do not wish to discuss further the possible representations of the vibration process, before more complicated cases have been calculated successfully from the new stand-point. It is not decided that the results will merely re-echo those of the usual quantum theory. For example, if the relativistic Kepler problem be worked out, it is found to lead in a remarkable manner to *half-integral partial* quanta (radial and azimuthal).

Still, a few remarks on the representation of the vibration may be permitted. Above all, I wish to mention that I was led to these deliberations in the first place by the suggestive papers of M. Louis de Broglie,[6] and by reflecting over the space distribution of those "phase waves," of which he

---

[6]L. de Broglie, *Ann. de Physique* (10) 3, p. 22, 1925. (Thèses, Paris, 1924.)

has shown that there is always a *whole number*, measured along the path, present on each period or quasi-period of the electron. The main difference is that de Broglie thinks of progressive waves, while we are led to stationary proper vibrations if we interpret our formulae as representing vibrations. I have lately shown[7] that the Einstein gas theory can be based on the consideration of such stationary proper vibrations, to which the dispersion law of de Broglie's phase waves has been applied. The above reflections on the atom could have been represented as a generalisation from those on the gas model.

If we take the separate functions (18), multiplied by a surface harmonic of order n, as the description of proper vibration processes, then the quantity E must have something to do with the related *frequency*. Now in vibration problems we are accustomed to the "parameter" (usually called λ) being proportional to the *square* of the frequency. However, in the first place, such a statement in our case would lead to *imaginary* frequencies for the *negative* E-values, and, secondly, instinct leads us to believe that the energy must be proportional to the frequency itself and not to its square.

The contradiction is explained thus. There has been *no natural zero level* laid down for the "parameter" E of the variation equation (5), especially as the unknown function ψ appears multiplied by a function of r, which can be changed by a constant to meet a corresponding change in the zero level of E. Consequently, we have to correct our anticipations, in that not E itself – continuing to use the same terminology – but E increased by a certain constant is to be expected to be proportional to the square of the frequency. Let this constant be now *very great* compared with all the admissible negative E-values (which are already limited by (15)). Then firstly, the frequencies will become *real*, and secondly, since our E-values correspond to only relatively small frequency

---

[7] *Physik. Ztschr.* 27, p. 95, 1926.

differences, they will actually be very approximately proportional to these frequency differences. This, again, is all that our "quantum-instinct" can require, as long as the zero level of *energy* is not fixed.

The view that the frequency of the vibration process is given by

$$\nu = C'\sqrt{C + E} = C'\sqrt{C} + \frac{C'}{2\sqrt{C}}E \dots, \qquad (22)$$

where C is a constant very great compared with all the E's, has still another very appreciable advantage. It *permits an understanding of the Bohr frequency condition*. According to the latter the emission frequencies are proportional to the E-*differences*, and therefore from (22) also to the differences of the proper frequencies ν of those hypothetical vibration processes. But these proper frequencies are all very great compared with the emission frequencies, and they agree very closely among themselves. The emission frequencies appear therefore as deep "difference tones" of the proper vibrations themselves. It is quite conceivable that on the transition of energy from one to another of the normal vibrations, *something* – I mean the light wave – with a *frequency* allied to each frequency *difference*, should make its appearance. One only needs to imagine that the light wave is causally related to the *beats*, which necessarily arise at each point of space during the transition; and that the frequency of the light is defined by the number of times per second the intensity maximum of the beat-process repeats itself.

It may be objected that these conclusions are based on the relation (22), in its *approximate* form (after expansion of the square root), from which the Bohr frequency condition itself seems to obtain the nature of an approximation. This, however, is merely apparently so, and it is wholly avoided when the relativistic theory is developed and makes a profounder insight possible. The large constant C is naturally very intimately connected with the rest-energy of the electron (mc^2). Also the seemingly *new* and *independent* introduction of the constant h (already brought in by (20)), into the frequency condition, is cleared up, or rather avoided, by the relativistic theory. But unfortunately the correct establishment of the latter meets right away with certain difficulties, which have been already alluded to.

It is hardly necessary to emphasize how much more congenial it would be to imagine that at a quantum transition the energy changes over from one form of vibration to another, than to think of a jumping electron. The changing of the vibration form can take place continuously in space and time, and it can readily last as long as the emission process lasts empirically (experiments on canal rays by W. Wien); nevertheless, if during this transition the atom is placed for a comparatively short time in an electric field which alters the proper frequencies, then the beat frequencies are immediately changed sympathetically, and for just as long as the field operates. It is known that this experimentally established fact has hitherto presented the greatest difficulties. See the well-known attempt at a solution by Bohr, Kramers, and Slater.

Let us not forget, however, in our gratification over our progress in these matters, that the idea of only *one* proper vibration being excited whenever the atom does not radiate – if we must hold fast to this idea – is very far removed from the natural picture of a vibrating system. We know that a macroscopic system does not behave like that, but yields in general a *pot-pourri* of its proper vibrations. But we should not make up our minds too quickly on this point. A *pot-pourri* of proper vibrations would also be permissible for a single atom, since thereby no beat frequencies could arise other than those which, according to experience, the atom is capable of emitting *occasionally*. The actual sending out of many of these spectral lines simultaneously by the same atom does not contradict experience. It is thus conceivable that only in the normal state (and approximately in certain "meta-stable" states) the atom vibrates with one proper frequency and just for this reason does *not* radiate, namely, because no beats arise. The *stimulation* may consist of a simultaneous excitation of one or of several other proper frequencies, whereby beats originate and evoke emission of light.

Under all circumstances, I believe, the proper functions, which belong to the *same* frequency, are in general all simultaneously stimulated. Multipleness of the proper values corresponds, namely, in the language of the previous theory to *degeneration.* To the reduction of the quantisation of degenerate systems probably corresponds the arbitrary partition of the energy among the functions belonging to *one* proper value.

Addition at the proof correction on 28.2.1926.

In the case of conservative systems in classical mechanics, the variation problem can be formulated in a neater way than was previously shown, and without express reference to the Hamilton-Jacobi differential equation. Thus, let $T(q, p)$ be the kinetic energy, expressed as a function of the coordinates and momenta, V the potential energy, and $d\tau$ the volume element of the space, "measured rationally," i.e. it is not simply the product $dq_1 \, dq_2 \, dq_3 \ldots dq_n$, but this divided by the square root of the discriminant of the quadratic form $T(q, p)$. (Cf. Gibbs' *Statistical Mechanics.*) Then let ψ be such as to make the "Hamilton integral"

$$\int d\tau \left\{ K^2 T \left(q, \frac{\partial \psi}{\partial q} \right) + \psi^2 V \right\} \tag{23}$$

stationary, while fulfilling the *normalising, accessory condition*

$$\int \psi^2 d\tau = 1. \tag{24}$$

24

The proper values of this variation problem are then *the stationary values* of integral (23) and yield, according to our thesis, *the quantum-levels of the energy.*

It is to be remarked that in the quantity α_2 of (14'') we have essentially the well-known Sommerfeld expression $-\frac{B}{\sqrt{A}} + \sqrt{C}$. (Cf. *Atombau*, 4th (German) ed., p. 775.)

Physical Institute of the University of Zürich.
(Received January 27, 1926.)

QUANTISATION AS A PROBLEM OF PROPER VALUES (PART II)

(*Annalen der Physik* (4), vol. 79, 1926)

§ 1. The Hamiltonian Analogy between Mechanics and Optics

Before we go on to consider the problem of proper values for further special systems, let us throw more light on the *general* correspondence which exists between the Hamilton-Jacobi differential equation of a mechanical problem and the "allied" *wave equation*, i.e. equation (5) of Part I. in the case of the Kepler problem. So far we have only briefly described this correspondence on its external analytical side by the transformation (2), which is in itself unintelligible, and by the equally incomprehensible transition from the *equating to zero* of a certain expression to the postulation that the *space integral* of the said expression shall be *stationary*.[1]

The inner connection between Hamilton's theory and the process of wave propagation is anything but a new idea. It was not only well known to Hamilton, but it also served him as the

---

[1] This procedure will *not be pursued further* in the present paper. It was only intended to give a provisional, quick survey of the external connection between the wave equation and the Hamilton-Jacobi equation. ψ is not actually the action function of a definite motion in the relation stated in (2) of Part I. On the other hand the connection between the wave equation and the variation problem is of course very real; the integrand of the stationary integral is the Lagrange function for the wave process.

starting-point for his theory of mechanics, which grew[2] out of his *Optics of Non-homogeneous Media*. Hamilton's variation principle can be shown to correspond to Fermat's *Principle* for a wave propagation in con- figuration space (q-space), and the Hamilton-Jacobi equation expresses Huygens' *Principle* for this wave propagation. Unfortunately this powerful and momentous conception of Hamilton is deprived, in most modern reproductions, of its beautiful raiment as a superfluous accessory, in favour of a more colourless representation of the analytical correspondence.[3]

Let us consider the general problem of conservative systems in classical mechanics. The Hamilton-Jacobi equation runs

$$\frac{\partial W}{\partial t} + T\left(q_k, \frac{\partial W}{\partial q_k}\right) + V(q_k) = 0. \tag{1}$$

W is the action function, i.e. the time integral of the Lagrange function $T - V$ along a path of the system as a function of the end points and the time, q_k is a representative position coordinate; T is the kinetic energy as function of the q's and momenta, being a quadratic form of the latter, for which, as prescribed, the partial derivatives of W with respect to the q's are written. V is the potential energy. To solve the equation put

$$W = -Et + S(q_k), \tag{2}$$

---

[2]Cf. e.g. E. T. Whittaker's *Anal. Dynamics*, chap. xi.

[3]Felix Klein has since 1891 repeatedly developed the theory of Jacobi from quasi-optical considerations in non-Euclidean higher space in his lectures on mechanics. Cf. F. Klein, *Jahresber. d. Deutsch. Math. Ver* 1, 1891, and *Zeits. f. Math. u. Phys.* 46, 1901 (*Ges.-Abh.* ii. pp. 601 and 603). In the second note, Klein remarks reproachfully that his discourse at Halle ten years previously, in which he had discussed this correspondence and emphasized the great significance of Hamilton's optical works, had "not obtained the general attention, which he had expected." For this allusion to F. Klein, I am indebted to a friendly communication from Prof. Sommerfeld. See also *Atombau*, 4th ed., p. 803.

and obtain

$$2T\left(q_k, \frac{\partial W}{\partial q_k}\right) = 2(E - V). \qquad (1')$$

E is an arbitrary integration constant and signifies, as is known, the energy of the system. Contrary to the usual practice, we have let the function W remain itself in $(1')$, instead of introducing the time-free function of the coordinates, S. That is a mere superficiality.

Equation $(1')$ can now be very simply expressed if we make use of the method of Heinrich Hertz. It becomes, like all geometrical assertions in configuration space (space of the variables q_k), especially simple and clear if we introduce into this space a non-Euclidean metric by means of the kinetic energy of the system.

Let T be the kinetic energy as function of the velocities q_k, not of the momenta as above, and let us put for the line element

$$ds^2 = 2T(q_k, \dot{q}_k)dt^2. \qquad (3)$$

The right-hand side now contains dt only externally and represents (since $\dot{q}_k dt = dq_k$) a quadratic form of the dq_k's.

After this stipulation, conceptions such as angle between two line elements, perpendicularity, divergence and curl of a vector, gradient of a scalar, Laplacian operation ($=$ div grad) of a scalar, and others, may be used in the same simple way as in three-dimensional Euclidean space, and we may use in our thinking the Euclidean three-dimensional representation with impunity, except that the analytical expressions for these ideas become a very little more complicated, as the line element (3) must everywhere replace the Euclidean line element. *We stipulate, that in what follows, all geometrical statements in q-space are to be taken in this non-Euclidean sense.*

One of the most important modifications for the calculation is that we must distinguish carefully between covariant and contravariant components of a vector or tensor. But this complication is not any greater than that which occurs in the case of an oblique set of Cartesian axes.

The dq_k's are the prototype of a contra variant vector. The coefficients of the form $2T$, which depend on the dq_k's, are therefore of a covariant character and form the covariant fundamental tensor. $2T$ is the contravariant form belonging to $2T$ because the momenta are known to form the covariant vector belonging to the speed vector $\dot{q}_k$, the momentum being the velocity vector in covariant form. The left side of (1′) is now simply the contravariant fundamental form, in which the $\frac{\partial W}{\partial q_k}$'s are brought in as variables. The latter form the components of the vector, – according to its nature covariant,

$$\text{grad W.}$$

(The expressing of the kinetic energy in terms of momenta instead of speeds has then *this* significance, that covariant vector components can only be introduced in a contravariant form if something intelligible, i.e. invariant, is to result.)

Equation (1′) is equivalent thus to the simple statement

$$(\text{grad W})^2 = 2(\text{E} - \text{V}), \tag{1″}$$

or

$$|\text{grad W}| = \sqrt{2(E - V)}. \tag{1‴}$$

This requirement is easily analysed. Suppose that a function W, of the form (2), has been found, which satisfies it. Then this function can be clearly represented for every definite t, if the family of surfaces $W = $ const be described in q-space and to each member a value of W be ascribed.

Now, on the one hand, as will be shown immediately, equation (1‴) gives an exact rule for constructing all the other surfaces of the family and obtaining their W-values from any single member, *if the latter and its W-value is known.* On the other hand, if the sole necessary data for the construction, viz. *one* surface and its W-value be *given quite arbitrarily*, then from the rule, which presents just *two* alternatives, there may be completed one of the functions W fulfilling the given requirement. Provisionally, the time is regarded as constant. – The construction rule therefore *exhausts* the contents of the differential equation; *each* of its solutions can be obtained from a suitably chosen surface and W-value.

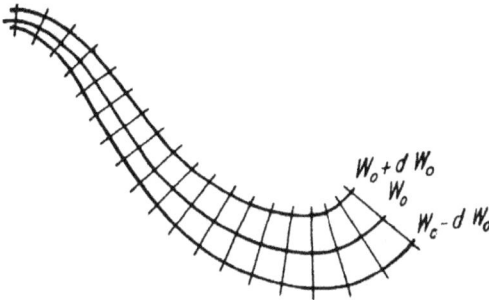

$W_0 + d\,W_0$
W_0
$W_c - d\,W_0$

Fig. 1

Let us consider the construction rule. Let the value W_0 be given in Fig. 1 to an arbitrary surface. In order to find the surface $W_0 + dW_0$, take *either* side of the given surface as the positive one, erect the normal at each point of it and cut off (with due regard to the sign of dW_0) the *step*

$$ds = \frac{dW_0}{\sqrt{2(E - V)}}. \tag{4}$$

The locus of the end points of the steps is the surface $W_0 + dW_0$. Similarly, the family of surfaces may be constructed successively on both sides.

The construction has a *double* interpretation, as the *other* side of the given surface might have been taken as positive for the first step). This ambiguity does not hold for later steps, i.e. at any later stage of the process we cannot change arbitrarily the sign of the sides of the surface, at which we have arrived, as this would involve in general a discontinuity in the first differential coefficient of W. Moreover, the two families obtained in the two cases are clearly identical; the W-values merely run in the opposite direction.

Let us consider now the very simple dependence on the *time*. For this, (2) shows that at any later (or earlier) instant $t+t'$ the same group of surfaces illustrates the W-distribution, though different If -values are associated with the individual members, namely, from each W-value ascribed at time t there must be subtracted Et'. The W-values wander, as it were, from surface to surface according to a definite, simple law, and for positive E in the direction of W increasing. Instead of this, however, we may imagine that the *surfaces* wander in such a way that each of them continually takes the place and exact form of the following one, and always carries its W-value *with* it. The rule for this wandering is given by the fact that the surface W_0 at time $t+dt$ must have reached that place, which at t was occupied by the surface $W_0 + Edt$. This will be attained according to (4), if each point of the surface W_0 is allowed to move in the direction of the positive normal through a distance

$$ds = \frac{Edt}{\sqrt{2(E - V)}}. \tag{5}$$

That is, the surfaces move with a *normal velocity*

$$u = \frac{ds}{dt} = \frac{E}{\sqrt{2(E - V)}}, \tag{6}$$

which, when the constant E is given, is a pure function of position.

Now it is seen that our system of surfaces W=const, can be conceived as the system of wave surfaces of a progressive but stationary wave motion in q-space, for which the value of the phase velocity at every point in the space is given by (6). For the normal construction can clearly bo replaced by the construction of elementary Huygens waves (with radius

(5)), and then of their envelope. The "index of refraction" is proportional to the reciprocal of (6), and is dependent on the position but not on the direction. The q-space is thus optically non-homogeneous but is isotropic. The elementary waves are "spheres", though of course – let me repeat it expressly once more – in the sense of the line-element (3).

The function of action W plays the part of the phase of our wave system. The Hamilton-Jacobi equation is the expression of Huygens' principle. If, now, Fermat's principle be formulated thus,

$$0 = \delta \int_{P_1}^{P_2} \frac{ds}{u} = \delta \int_{P_1}^{P_2} \frac{ds\sqrt{2(E-V)}}{E}$$
$$= \delta \int_{t_1}^{t_2} \frac{2T}{E} dt = \frac{1}{E} \delta \int_{t_1}^{t_2} 2T dt, \tag{7}$$

we are led directly to Hamilton's principle in the form given by Maupertuis (where the time integral is to be taken with the usual grain of salt, i.e. $T+V = E =$ constant, even during the variation). The "rays", i.e . the orthogonal trajectories of the wave surfaces, are therefore the *paths* of the system for the value E of the energy, in agreement with the well-known system of equations

$$p_k = \frac{\partial W}{\partial q_k} \tag{8}$$

which states, that a set of system paths can be derived from each special function of action, just like a fluid motion from its velocity potential.[4] (The momenta p_k form the covariant velocity vector, which equations (8) assert to be equal to the gradient of the function of action.)

---

[4] See especially A. Einstein, *Verh. d. D. Physik. Ges.* 19, pp. 77, 82, 1917. The framing of the quantum conditions here is the most akin, out of all the older attempts, to the present one. De Broglie has returned to it.

Although in these deliberations on wave surfaces we speak of velocity of propagation and Huygens' principle, we must regard the analogy as one between mechanics and *geometrical* optics, and not physical or *undulatory* optics. For the idea of "rays", which is the essential feature in the mechanical analogy, belongs to *geometrical* optics; it is only clearly defined in the latter. Also Fermat's principle can be applied in geometrical optics without going beyond the idea of index of refraction. And the system of W-surfaces, regarded as wave surfaces, stands in a somewhat looser relationship to mechanical motion, inasmuch as the image point of the mechanical system in no wise moves along the ray with the wave velocity u, but, on the contrary, its velocity (for constant E) is proportional to $\frac{1}{u}$. It is given directly from (3) as

$$v = \frac{ds}{dt} = \sqrt{2T} = \sqrt{2(E - V)}. \tag{9}$$

This non-agreement is obvious. Firstly, according to (8), the system's point velocity is *great* when grad W is great, i.e. where the W-surfaces are closely crowded together, i.e. where u is small. Secondly, from the definition of W as the time integral of the Lagrange function, W *alters* during the motion (by $(T - V)dt$ in the time dt), and so the image point *cannot* remain continuously in contact with the same W-surface.

And important ideas in wave theory, such as amplitude, wave length, and frequency – or, speaking more generally, the wave *form* – do not enter into the analogy at all, as there exists no mechanical parallel; even of the wave function itself there is no mention beyond that W has the meaning of the *phase* of the waves (and this is somewhat hazy owing to the wave *form* being undefined).

If we find in the whole parallel merely a satisfactory means of contemplation, then this defect is not disturbing, and we would regard any attempt to supply it as idle trifling, believing the analogy to be precisely with *geometrical*, or at furthest, with a very primitive form of wave optics, and not with the fully developed undulatory optics. That geometrical optics is only a rough approximation for *Light* makes no difference. To *preserve* the analogy on the further development of the optics of q-space on the lines of wave theory, we must take good care not to depart markedly from the limiting case of geometrical optics, i.e. must choose[5] the *wave length* sufficiently small, i.e. small compared with all the path dimensions. Then the additions do not teach anything new; the picture is only draped with superfluous ornaments.

So we might think to begin with. But even the first attempt at the development of the analogy to the wave theory leads to such striking results, that a quite different suspicion arises: *we know today, in fact, that our classical mechanics fails for very small dimensions of the path and for very great curvatures.* Perhaps this failure is in strict analogy with the failure of geometrical optics, i.e. "the optics of infinitely small wave lengths", that becomes evident as soon as the obstacles or apertures are no longer great compared with the real, finite, wave length. Perhaps our classical mechanics is the *complete* analogy of geometrical optics and as such is wrong and not in agreement with reality; it fails whenever the radii of curvature and dimensions of the path are no longer great compared with a certain wave length, to which, in q-space, a real meaning is attached. Then it becomes a question of searching[6] for an undulatory mechanics, and the most obvious way is the working out of the Hamiltonian analogy on the lines of undulatory optics.

---

[5]Cf. for the optical case, A. Sommerfeld and Iris Runge, *Ann. d. Phys.* 35, p. 290, 1911. There (in the working out of an oral remark of P. Debye), it is shown, how the equation of *first* order and second *degree* for the *phase* ("Hamiltonian equation") may be accurately derived from the equation of the *second* order and *first* degree for the *wave function* ("wave equation"), in the limiting case of vanishing wave length.

[6]Cf. A. Einstein, *Berl. Ber.* p. 9 *et seq.* 1925.

§ 2. "Geometrical" and "Undulatory" Mechanics

We will at first assume that it is fair, in extending the analogy, to imagine the above-mentioned wave system as consisting of *sine* waves. This is the simplest and most obvious case, yet the *arbitrariness*, which arises from the *fundamental significance* of this assumption, must be emphasized. The wave function has thus only to contain the time in the form of a factor, sin(...), where the argument is a linear function of W. The coefficient of W must have the dimensions of the reciprocal of action, since W has those of action and the phase of a sine has zero dimensions. We assume that it is quite universal, i.e. that it is not only independent of E, but also of the nature of the mechanical system. We may then at once denote it by $\frac{2\pi}{h}$. The time factor then is

$$\sin\left(\frac{2\pi W}{h} + \text{const.}\right)$$
$$= \sin\left(-\frac{2\pi Et}{h} + \frac{2\pi S(q_k)}{h} + \text{const.}\right) \tag{10}$$

Hence the *frequency* ν of the waves is given by

$$\nu = \frac{E}{h}. \tag{11}$$

Thus we get the frequency of the -space waves to be proportional to the energy of the system, in a manner which is not markedly artificial.[7] This is only true of course if E is absolute and not, as in classical mechanics, indefinite to the extent of an additive constant. By (6) and (11) the *wave length* is *independent* of this additive constant, being

$$\lambda = \frac{u}{v} = \frac{h}{\sqrt{2(E-V)}}, \tag{12}$$

---

[7]In Part I. this appeared merely as an approximate equation, derived from a pure speculation.

and we know the term under the root to be double the kinetic energy. Let us make a preliminary rough comparison of this wave length with the dimensions of the orbit of a hydrogen electron as given by classical mechanics, taking care to notice that a "step" in q-space has not the dimensions of length, but length multiplied by the square root of mass, in consequence of (3). A has similar dimensions. We have therefore to divide A by the dimension of the orbit, a cm., say, and by the square root of m, the mass of the electron. The quotient is of the order of magnitude of

$$\frac{h}{mva},$$

where v represents for the moment the electron's velocity (cm./sec.). The denominator mva is of the order of the mechanical moment of momentum, and this is at least of the order of 10^{-27} for Kepler orbits, as can be calculated from the values of electronic charge and mass independently of all quantum theories. We thus obtain the correct order for the *limit of the approximate region of validity of classical mechanics*, if we identify our constant h with Planck's quantum of action – and this is only a preliminary attempt.

If in (6), E is expressed by means of (11) in terms of ν, then we obtain

$$u = \frac{h\nu}{\sqrt{2(h\nu - V)}}. \tag{6'}$$

The dependence of the wave velocity on the energy thus becomes a particular kind of dependence on the *frequency*, i.e. it becomes a law of *dispersion* for the waves. This law is of great interest. We have shown in § 1 that the wandering wave surfaces are only loosely connected with the motion of the system point, since their velocities are not equal and cannot be equal. According to (9), (11), and (6′) the system's velocity v has thus also a concrete significance for the wave. We

36

verify at once that

$$v = \frac{d\nu}{d\left(\frac{\nu}{u}\right)},\qquad(13)$$

i.e. the velocity of the system point is that of a *group of waves*, included within a small range of frequencies (signal-velocity). We find here again a theorem for the "phase waves" of the electron, which M. de Broglie had derived, with essential reference to the relativity theory, in those fine researches,[8] to which I owe the inspiration for this work. We see that the theorem in question is of wide generality, and does not arise solely from relativity theory, but is valid for every conservative system of ordinary mechanics.

We can utilise this fact to institute a much more innate connection between wave propagation and the movement of the representative point than was possible before. We can attempt to build up a wave group which will have relatively small dimensions in every direction. Such a wave group will then presumably obey the same laws of motion as a single image point of the mechanical system. It will then give, so to speak, an *equivalent* of the image point, so long as we can look on it as being approximately confined to a point, i.e . so long as we can neglect any spreading out in comparison with the dimensions of the path of the system. This will only be the case when the path dimensions, and especially the radius of curvature of the path, are very great compared with the wave length. For, in analogy with ordinary optics, it is obvious from what has been said that not only must the dimensions of the wave group not be reduced below the order of magnitude of the wave length, but, on the contrary, the group must extend in all directions over a large number of wave lengths, if it is to be *approximately monochromatic*. This, however, must be postulated, since the wave group must move about as a whole

---

[8] L. de Broglie, *Ann. de Physique* (10) 3, p. 22, 1925. (Thèses, Paris, 1924.)

with a definite group velocity and correspond to a mechanical system of *definite energy* (cf. equation 11).

So far as I see, such groups of waves can be constructed on exactly the same principle as that used by Debye[9] and von Laue[10] to solve the problem in ordinary optics of giving an exact analytical representation of a cone of rays or of a sheaf of rays. From this there comes a very interesting relation to that part of the Hamilton-Jacobi theory not described in § 1, viz. the well-known derivation of the equations of motion in integrated form, by the differentiation of a complete integral of the Hamilton-Jacobi equation with respect to the constants of integration. As we will see immediately, the system of equations called after Jacobi is equivalent to the statement: the image point of the mechanical system continuously corresponds to *that* point, where a certain continuum of wave trains coalesces in *equal phase*.

In optics, the representation (strictly on the wave theory) of a "sheaf of rays" with a sharply defined finite cross-section, which proceeds to a focus and then diverges again, is thus carried out by Debye. A *continuum* of *plane* wave trains, each of which alone would fill the *whole* space, is superposed. The *continuum* is produced by letting the wave normal vary throughout the given solid angle. The waves then destroy one another almost completely by interference outside a certain double cone; they represent exactly, on the wave theory, the desired limited sheaf of rays and also the diffraction phenomena, necessarily occasioned by the limitation. We can represent in this manner an *infinitesimal* cone of rays just as well as a finite one, if we allow the wave normal of the group to vary only inside an infinitesimal solid angle. This has been utilised by von Laue in his famous paper on the degrees of freedom of a sheaf of rays.[11] Finally, instead of working with

---

[9]P. Debye, *Ann. d. Phys.* 30, p. 755, 1909.

[10]M. v. Laue, *idem* 44, p. 1197 (§ 2), 1914.

[11]*Loc. cit.*

waves, hitherto tacitly accepted as purely monochromatic, we can also allow the *frequency* to vary within an infinitesimal interval, and by a suitable distribution of the amplitudes and phases can confine the disturbance to a region which is relatively small in the longitudinal direction also. So we succeed in representing analytically a "parcel of energy" of relatively small dimensions, which travels with the speed of light, or when dispersion occurs, with the group velocity. Thereby is given the instantaneous *position* of the parcel of energy – if the detailed structure is not in question – in a very plausible way as that point of space where *all* the superposed plane waves meet in *exactly* agreeing phase.

We will now apply these considerations to the q-space waves. We select, at a definite time t, a definite point P of q-space, through which the parcel of waves passes in a given direction R at that time. In addition let the mean frequency ν or the mean E-value for the packet be also given. These conditions correspond exactly to postulating that at a given time the mechanical system is starting from a given configuration with given velocity components. (Energy *plus* direction is equivalent to velocity components.)

In order to carry over the optical construction, we require firstly *one* set of wave surfaces with the desired frequency, i.e. *one* solution of the Hamilton-Jacobi equation (1′) for the given E-value. This solution, W, say, is to have the following property: the surface of the set which passes through P at time t, which we may denote bv

$$W = W_0, \qquad (14)$$

must have its normal at P in the prescribed direction R. But this is still not enough. We must be able to vary to an infinitely small extent this set of waves W in an n-fold manner (n = number of degrees of freedom), so that the wave normal will sweep out an infinitely small $(n-1)$ dimensional space angle at the point P, and so that the frequency $\frac{E}{h}$ will vary in an infinitely small *one*-dimensional region, whereby care is taken that all members of the infinitely small n-dimensional continuum of sets of waves meet together at time t in the point P in exactly agreeing phase. Then it is a question of finding at any other time *where* that point lies at which this agreement of phases occurs.

To do this, it will be sufficient if we have at our disposal a solution W of the Hamilton-Jacobi equation, which is dependent not only on the constant E, here denoted by α_1 but also on $(n-1)$ additional constants $\alpha_2, \alpha_3 \ldots \alpha_n$, in such a way that it cannot be written as a function of less than n combinations of these n constants. For then we can, firstly, bestow on α_1 the value prescribed for E, and, secondly, define $\alpha_2, \alpha_3 \ldots \alpha_n$, so that the surface of the set passing through the point P has at P the prescribed normal direction. Henceforth we understand by $\alpha_1, \alpha_2 \ldots \alpha_n$, *these* values, and take (14) as the surface of *this* set, which passes through the point P at time t. Then we consider the *continuum of sets* which belongs to the α_k-values of an adjacent infinitesimal α_k-region. A member of this continuum, i.e. therefore *a set*, will be given by

$$W + \frac{\partial W}{\partial \alpha_1} d\alpha_1 + \frac{\partial W}{\partial \alpha_2} d\alpha_2 + \ldots + \frac{\partial W}{\partial \alpha_n} d\alpha_n = \text{const.} \quad (15)$$

for a *fixed* set of values of $d\alpha_1, d\alpha_2 \ldots d\alpha_n$, and varying constant. That member of *this* set, i.e. therefore that single surface, which goes through P at time t will be defined by the following choice of the const.,

$$W + \frac{\partial W}{\partial \alpha_1} d\alpha_1 + \ldots + \frac{\partial W}{\partial \alpha_n} d\alpha_n$$
$$= W_0 + \left(\frac{\partial W}{\partial \alpha_1} \right)_0 d\alpha_1 + \ldots + \left(\frac{\partial W}{\partial \alpha_n} \right)_0 d\alpha_n, \quad (15')$$

where $\left(\frac{\partial W}{\partial \alpha_1} \right)_0$, etc., are the *constants* obtained by substituting in the differential coefficients the coordinates of the point

40

P and the value t of the time (which latter really only occurs in $\frac{\partial W}{\partial \alpha_1}$).

The surfaces (15′) for all possible sets of values of $d\alpha_1, d\alpha_2 \ldots d\alpha_n$ form on their part α set. They all go through the point P at time t their wave normals continuously sweep out a little $(n-1)$ dimensional solid angle and, moreover, their E-parameter also varies within a small region. The set of surfaces (15′) is so formed that each of the sets (15) supplies *one* representative to (15′), namely, that member which passes through P at time t.

We will now assume that the phase angles of the wave functions which belong to the sets (15) happen to agree precisely for those representatives which enter the set (15′). They agree therefore at time t at the point P.

We now ask: Is there, at *any arbitrary time*, a point where all surfaces of the set (15′) cut one another, *and in which, therefore*, all the wave functions which belong to the sets (15) agree in phase? The answer is: *There exists* a point of agreeing phase, but it is not the common intersection of the surfaces of set (15′), for such does not exist at any subsequent arbitrary time. Moreover, the point of phase agreement arises in *such a way* that the sets (15) *continuously exchange* their representatives given to (15′).

That is shown thus. There must hold

$$W = W_0,$$
$$\frac{\partial W}{\partial \alpha_1} = \left(\frac{\partial W}{\partial \alpha_1} \right)_0,$$
$$\frac{\partial W}{\partial \alpha_2} = \left(\frac{\partial W}{\partial \alpha_2} \right)_0, \quad (16)$$
$$\ldots$$
$$\frac{\partial W}{\partial \alpha_n} = \left(\frac{\partial W}{\partial \alpha_n} \right)_0,$$

simultaneously for the common meeting point of all members

of (15') at any time, because the $d\alpha_1$'s are arbitrary within a small region. In these $n + 1$ equations, the right-hand sides are constants, and the left are functions of the $n + 1$ quantities $q_1, q_2, \ldots q_n, t$. The equations are satisfied by the initial system of values, i.e. by the coordinates of P and the initial time t. For another arbitrary value of t, they will have *no* solutions in $q_1, \ldots q_n, t$, but will *more than define* the system of these n quantities.

We may proceed, however, as follows. Let us leave the first equation, $W = W_0$, aside at first, and define the q'_ks as functions of the time and the constants according to the remaining n equations. Let this point be called Q. By it, naturally, the *first* equation will *not* be satisfied, but the left-hand side will differ from the right by a certain value. If we go back to the derivation of system (16) from (15'), what we have just said means that though Q is not a common point for the set of surfaces (15'), it is so, however, for a set which results from (15'), if we alter the right-hand side of equation (15') by an amount which is constant for all the surfaces. Let this new set be (15''). For it, therefore, Q is a common point. The new set results from (15'), as stated above, by an exchange of the representatives in (15'). This exchange is occasioned by the alteration of the constant in (15), *by the same amount*, for all representatives. Hence the *phase angle* is altered by the same amount for all representatives. The new representatives, i.e. the members of the set we have called (15''), which meet in the point Q, agree in phase angle just as the old ones did. This amounts therefore to saying:

The point Q which is defined as a function of the time by the n equations

$$\frac{\partial W}{\partial \alpha_1} = \left(\frac{\partial W}{\partial \alpha_1}\right)_0, \ldots, \frac{\partial W}{\partial \alpha_n} = \left(\frac{\partial W}{\partial \alpha_n}\right)_0, \qquad (17)$$

continues to be a point of agreeing phase for the whole aggregate of wave sets (15).

Of all the n-surfaces, of which Q is shown by (17) to be the common point, only the first is variable; the others remain fixed (only the first of equations (17) contains the time). The $n - 1$ fixed surfaces determine the path of the point Q as their line of intersection. It is easily shown that this line is the orthogonal trajectory of the set $W = $ const. For, by hypothesis, W satisfies the Hamilton-Jacobi equation (1') identically in $\alpha_2, \alpha_3 \ldots \alpha_n$. If we now differentiate the Hamilton-Jacobi equation with respect to α_k ($k = 2, 3, \ldots n$), we get the statement that the normal to a surface, $\frac{\partial W}{\partial \alpha_k} = $ const. is *perpendicular*, at every point on it, to the normal of the surface, $W = $ const., which passes through that point, i.e. that each of the two surfaces *contains* the normal to the other. If the line of intersection of the $n - 1$ fixed surfaces (17) has no branches, as is generally the case, then must each line element of the intersection, as the sole common line element of the $n - 1$ surfaces, coincide with the normal of the W-surface, passing through the same point, i.e. the line of intersection is the orthogonal trajectory of the W-surfaces. Q.E.D.

We may sum up the somewhat detailed discussion, which has led us to equations (17), in a much shorter or (so to speak) shorthand fashion, as follows: W denotes, apart from a universal constant ($\frac{1}{h}$), the phase angle of the wave function. If we now deal not merely with *one*, but with a continuous manifold of wave systems, and if these are continuously arranged by means of any continuous parameters α_i, then the equations $\frac{\partial W}{\partial \alpha_i} = $ const. express the fact that all infinitely adjacent individuals (wave systems) of this manifold agree in phase. These equations therefore define the geometrical locus of the points of agreeing phase. If the equations are sufficient, this locus shrinks to one point; the equations then define *the point* of phase agreement as a function of the time.

Since the system of equations (17) agrees with the known second system of equations of Jacobi, we have thus shown:

The point of phase agreement for certain infinitesimal man-

ifolds of wave systems, containing n *parameters, moves according to the same laws as the image point of the mechanical system.*

I consider it a very difficult task to give an exact proof that the superposition of these wave systems really produces a noticeable disturbance in only a relatively small region surrounding the point of phase agreement, and that everywhere else they practically destroy one another through interference, or that the above statement turns out to be true at least for a suitable choice of the amplitudes, and possibly for a special choice of the *form* of the wave surfaces. I will advance the physical hypothesis, which I wish to attach to what is to be proved, without attempting the proof. The latter will only be worth while if the hypothesis stands the test of trial *and* if its application should *require* the exact proof.

On the other hand, we may be sure that the region to which the disturbance may be confined still contains in all directions a great number of wave lengths. This is directly evident, firstly, because so long as we are only a *few* wave lengths distant from the point of phase agreement, then the agreement of phase is hardly disturbed, as the interference is still ablest as favourable as it is at the point itself. Secondly, a glance at the three-dimensional Euclidean case of ordinary optics is sufficient to assure us of this general behaviour.

What I now categorically conjecture is the following:

The true mechanical process is realised or represented in a fitting way by the *wave processes* in q-space, and not by the motion of image points in this space. The study of the motion of image points, which is the object of classical mechanics, is only an approximate treatment, and has, as such, just as much justification as geometrical or "ray" optics has, compared with the true optical process. A macroscopic mechanical process will be portrayed as a wave signal of the kind described above, which can approximately enough be regarded as confined to a point compared with the geometrical structure of the path. We have seen that the same laws of motion hold exactly for such a signal or group of waves as are advanced by classical mechanics for the motion of the image point. This manner of treatment, however, loses all meaning where the structure of the path is no longer very large compared with the wave length or indeed is comparable with it. Then we *must* treat the matter strictly on the wave theory, i.e. we must proceed from the *wave equation* and not from the fundamental equations of mechanics, in order to form a picture of the manifold of the possible processes. These latter equations are just as useless for the elucidation of the micro-structure of mechanical processes as geometrical optics is for explaining the *phenomena of diffraction.*

Now that a certain interpretation of this micro-structure has been successfully obtained as an addition to classical mechanics, although admittedly under new and very artificial assumptions, an interpretation bringing with it practical successes of the highest importance, it seems to me very significant that these theories – I refer to the forms of quantum theory favoured by Sommerfeld, Schwarzschild, Epstein, and others – bear a very close relation to the Hamilton- Jacobi equation and the theory of its solution, i.e. to that form of classical mechanics which already points out most clearly the true undulatory character of mechanical processes. The Hamilton-Jacobi equation corresponds to Huygens' Principle (in its old simple form, not in the form due to Kirchhoff). And just as this, supplemented by some rules which are not intelligible in geometrical optics (Fresnel's construction of zones), can explain to a great extent the phenomena of diffraction, so light can be thrown on the processes in the atom by the theory of the action-function. But we inevitably became involved in irremovable contradictions if we tried, as was very natural, to maintain also the idea of *paths of systems* in these processes; just as we find the tracing of the course of a *light ray* to be meaningless, in the neighbourhood of a diffraction phenomenon.

We can argue as follows. I will, however, not yet give a conclusive picture of the actual process, which positively cannot be arrived at from this starting-point but only from an investigation of the wave equation; I will merely illustrate the matter qualitatively. Let us think of a wave group of the nature described above, which in some way gets into a small closed "path," whose dimensions are of the order of the wave length, and therefore *small* compared with the dimensions of the wave group itself. It is clear that then the "system path" in the sense of classical mechanics, i.e. the path of the point of exact phase agreement, will completely lose its prerogative, because there exists a whole continuum of points before, behind, and near the particular point, in which there is almost as complete phase agreement, and which describe totally different "paths." In other words, the wave group not only fills the whole path domain all at once but also stretches far beyond it in all directions.

In this sense do I interpret the "phase waves" which, according to de Broglie, accompany the path of the electron; in the sense, there- fore, that no special meaning is to be attached to the electronic path itself (at any rate, in the interior of the atom), and still less to the position of the electron on its path. And in this sense I explain the conviction, increasingly evident today, *firstly*, that real meaning has to be denied to the *phase* of electronic motions in the atom; *secondly*, that we can never assert that the electron at a definite instant is to be found on *any definite one* of the quantum paths, specialised by the quantum conditions; and *thirdly*, that the true laws of quantum mechanics do not consist of definite rules for the *single path*, but that in these laws the elements of the whole manifold of paths of a system are bound together by equations, so that apparently a certain reciprocal action exists between the different paths.[12]

It is not incomprehensible that a careful analysis of the experimentally known quantities should lead to assertions of this kind, if the experimentally known facts are the outcome of such a structure of the real process as is here represented. All these assertions systematically contribute to the relinquishing of the ideas of "place of the electron" and "path of the electron." If these are not given up, contradictions remain. This contradiction has been so strongly felt that it has even been doubted whether what goes on in the atom could ever be described within the scheme of space and time. From the philosophical standpoint, I would consider a conclusive decision in this sense as equivalent to a complete surrender. For we cannot really alter our manner of thinking in space and time, and what we cannot comprehend within it we cannot understand at all. There *are* such things – but I do not believe that atomic structure is one of them. From our standpoint, however, there is no reason for such doubt, although or rather *because* its appearance is extraordinarily comprehensible. So might a person versed in geometrical optics, after many attempts to explain diffraction phenomena by means of the idea of the ray (trustworthy for his macroscopic optics), which always came to nothing, at last think that the *Laws of Geometry* are not applicable to diffraction, since he continually finds that light rays, which he imagines as *rectilinear* and *independent* of each other, now suddenly show, even in homogeneous media, the most remarkable *curvatures*, and obviously *mutually influence* one another. I consider this analogy as *very* strict. Even for the unexplained *curvatures*, the analogy in the atom is not lacking – think of the "nonmechanical force," devised for the explanation of anomalous Zeeman effects.

---

[12]Cf. especially the papers of Heisenberg, Born, Jordan, and Dirac quoted later, and further N. Bohr, *Die Naturwissenschaften*, January 1926.

In what way now shall we have to proceed to the undulatory representation of mechanics for those cases where it is necessary? We must start, not from the fundamental equations of mechanics, but from a wave equation for q-space and consider the manifold of processes possible *according to it*. The wave equation has not been explicitly used or even put forward in this communication. The only datum for its construction is the *wave velocity*, which is given by (6) or (6′) as a function of the mechanical energy parameter or frequency respectively, and by this datum the wave equation is evidently not uniquely defined. It is not even decided that it must be definitely of the second order. Only the striving for simplicity leads us to try this to begin with. We will then say that for the wave function ψ we have

$$\operatorname{div}\operatorname{grad}\psi - \frac{1}{u^2}\ddot{\psi} = 0, \tag{18}$$

valid for all processes which only depend on the time through a factor $e^{2\pi i \nu t}$ Therefore, considering (6), (6′), and (11), we get, respectively,

$$\operatorname{div}\operatorname{grad}\psi + \frac{8\pi^2}{h^2}(h\nu - V)\psi = 0, \tag{18′}$$

and

$$\operatorname{div}\operatorname{grad}\psi + \frac{8\pi^2}{h^2}(E - V)\psi = 0. \tag{18″}$$

The differential operations are to be understood with regard to the line element (3). But even under the postulation of second order, the above is not the only equation consistent with (6). For it is possible to generalize by replacing $\operatorname{div}\operatorname{grad}\psi$ by

$$f(q_k)\operatorname{div}\left(\frac{1}{f(q_k)}\operatorname{grad}\psi\right), \tag{19}$$

where f may be an arbitrary function of the q's, which must depend in some plausible way on E, $V(q_k)$, and the coefficients of the line clement (8). (Think, e.g ., of $f = u$.) Our

48

postulation is again dictated by the striving for simplicity, yet I consider in this case that a wrong deduction is not out of the question.[13]

The substitution of a *partial* differential equation for the equations of dynamics in atomic problems appears at first sight a very doubtful procedure, on account of the multitude of solutions that such an equation possesses. Already classical dynamics had led not just to one solution but to a much too extensive manifold of solutions, viz. to a continuous set, while all experience seems to show that only a discrete number of these solutions is realised. The problem of the quantum theory, according to prevailing conceptions, is to select by means of the "quantum conditions" that discrete set of actual paths out of the continuous set of paths possible according to classical mechanics. It seems to be a bad beginning for a new attempt in this direction if the number of possible solutions has been *increased* rather than diminished.

It is true that the problem of classical dynamics also allows itself to be presented in the form of a *partial* equation, namely, the Hamilton-Jacobi equation. But the manifold of solutions of the problem does not correspond to the manifold of solutions of that equation. An arbitrary "complete" solution of the equation solves the mechanical problem *completely*; any *other* complete solution yields the same paths – they are only contained in another way in the manifold of paths.

Whatever the fear expressed about taking equation (18) as the foundation of atomic dynamics comes to, I will not positively assert that no further additional definitions will be required with it. But these will probably no longer be of such a completely strange and incomprehensible nature as the previous "quantum conditions," but will be of the type that we are accustomed to find in physics with a partial differential

---

[13]The introduction of $f(q_k)$ means that not only the "density" but also the "elasticity" varies with the position.

equation as initial or boundary conditions. They will be, in no way, analogous to the quantum conditions – because in all cases of classical dynamics, which I have investigated up till now, it turns out that equation (18) *carries within itself the quantum conditions*. It distinguishes in certain cases, and indeed in those where experience demands it, *of itself*, certain frequencies or energy levels as those which alone are possible for stationary processes, without any further assumption, other than the almost obvious demand that, as a physical quantity, the function ψ must be single-valued, finite, and continuous throughout configuration space.

Thus the fear expressed is transformed into its contrary, in any case in what concerns the energy levels, or let us say more prudently, the frequencies. (For the question of the "vibrational energy" stands by itself; we must not forget that it is only in the one electron problem that the interpretation as a vibration in real three-dimensional space is immediately suggested.) The definition of the quantum levels *no longer takes place* in two separated stages: (1) Definition of all paths dynamically possible. (2) *Discarding* of the greater part of those solutions and the selection of a few by special postulations; on the contrary, the quantum levels are *at once* defined as the proper values of equation (18), *which carries in itself its natural boundary conditions.*

As to how far an analytical simplification will be effected in this way in more complicated cases, 1 have not yet been able to decide. I) should, however, expect so. Most of the analytical investigators have the feeling that in the two-stage process, described above, there must be yielded in (1) the solution of a more complicated problem than is really necessary for the final result: energy as a (usually) very simple rational function of the quantum numbers. Already, as is known, the application of the Hamilton-Jacobi method creates a great simplification, as the actual calculation of the mechanical solution is avoided. It is sufficient to evaluate the integrals, which represent the momenta, merely for a closed complex path of integration instead of for a variable upper limit, and this gives much less trouble. Still the complete solution of the Hamilton-Jacobi equation must really be known, i.e. given by quadratures, so that the integration of the mechanical problem must in principle be effected for arbitrary initial values. In seeking for the proper values of a differential equation, we must usually, in practice, proceed thus. We seek the solution, firstly, with- out regard to boundary or continuity conditions, and from the form of the solution then pick out those values of the parameters, for which the solution satisfies the given conditions. Part I. supplies an example of this. We see by this example also, however – what is typical of proper value problems – that the solution was only given *generally* in an extremely inaccessible analytical form [equation (12) *loc. cit.*], but that it is extraordinarily simplified for those proper values belonging to the "natural boundary condition." I am not well enough informed to say whether *direct* methods have now been worked out for the calculation of the proper values. This is known to be so for the distribution of proper values of *high order.* But this limiting case is not of interest here; it corresponds to the classical, macroscopic mechanics. For spectroscopy and atomic physics, in general just the first 5 or 10 proper values will be of interest; even the *first alone* would be a great result – it defines the *ionisation potential.* From the idea, definitely outlined, that every problem of proper values allows itself to be treated as one of maxima and minima without direct reference to the differential equation, it appears to me very probable that direct methods will be found for the calculation, at least approximately, of the proper values, as soon as *urgent* need arises. At least it should be possible to test in individual cases whether the proper values, *known* numerically to all desired accuracy through spectroscopy, *satisfy* the problem or not.

I would not like to proceed without mentioning here that

at the present time a research is being prosecuted by Heisenberg, Born, Jordan, and other distinguished workers,[14] to remove the quantum difficulties, which has already yielded such noteworthy success that it cannot be doubted that it contains at least a part of the truth. In its *tendency*, Heisenberg's attempt stands very near the present one, as we have already mentioned. In its method, it is so totally different that I have not yet succeeded in finding the connecting link. I am distinctly hopeful that these two advances will not fight against one another, but on the contrary, just because of the extraordinary difference between the starting-points and between the methods, that they will supplement one another and that the one will make progress where the other fails. The strength of Heisenberg's programme lies in the fact that it promises to give the *line-intensities*, a question that we have not approached as yet. The strength of the present attempt – if I may be permitted to pronounce thereon – lies in the guiding, physical point of view, which creates a bridge between the macroscopic and microscopic mechanical processes, and which makes intelligible the outwardly different modes of treatment which they demand. For me, personally, there is a special charm in the conception, mentioned at the end of the previous part, of the emitted frequencies as "beats," which I believe will lead to an intuitive understanding of the intensity formulae.

§ 3. Application to Examples

We will now add a few more examples to the Kepler problem treated in Part I., but they will only be of the very simplest nature, since we have provisionally confined ourselves to

---

[14]W. Heisenberg, *Ztschr.f. Phys.* 33, p. 879, 1925; M. Born and P. Jordan, *ibid.* 34, p. 858, 1925; M. Born, W. Heisenberg, and P. Jordan, *ibid.* 35, p. 557, 1926; P. Dirac, *Proc. Roy. Soc.*, London, 109, p. 642, 1925.

classical mechanics, with no magnetic field.[15]

1. *The Planck Oscillator. The Question of Degeneracy*

Firstly we will consider the one-dimensional oscillator. Let the coordinate q be the displacement multiplied by the square root of the mass. The two forms of the kinetic energy then are

$$\bar{T} = \frac{1}{2}\dot{q}^2, \qquad T = \frac{1}{2}p^2. \tag{20}$$

The potential energy will be

$$V(q) = 2\pi^2 \nu_0^2 q^2, \tag{21}$$

where ν_0 is the proper frequency in the mechanical sense. Then equation (18) reads in this case

$$\frac{d^2\psi}{dq^2} + \frac{8\pi^2}{h^2}(E - 2\pi^2\nu_0^2 q^2)\psi = 0. \tag{22}$$

For brevity write

$$a = \frac{8\pi^2 E}{h^2}, \qquad b = \frac{16\pi^4\nu_0^2}{h^2}. \tag{23}$$

Therefore

$$\frac{d^2\psi}{dq^2} + (a - bq^2)\psi = 0. \tag{22'}$$

Introduce as independent variable

$$x = q\sqrt[4]{b}, \tag{24}$$

---

[15] In relativity mechanics and taking a magnetic field into account the statement of the Hamilton-Jacobi equation becomes more complicated. In the case of a single electron, it asserts that the *four-dimensional gradient* of the action function, *diminished* by a given vector (the four-potential), has a constant value. The translation of this statement into the language of the wave theory presents a good many difficulties.

and obtain

$$\frac{d^2\psi}{dx^2} + \left(\frac{a}{\sqrt{b}} - x^2\right)\psi = 0. \tag{22''}$$

The proper values and functions of this equation are *known*.[16] The proper values are, with the notation used here,

$$\frac{a}{\sqrt{b}} = 1, 3, 5, \ldots (2n+1), \ldots \tag{25}$$

The functions are the *orthogonal functions of Hermite*,

$$e^{-\frac{x^2}{2}} H_n(x). \tag{26}$$

$H_n(x)$ means the nth Hermite polynomial, which can be defined as

$$H_n(x) = (-1)^n e^{x^2} \frac{d^n e^{-x^2}}{dx^n}, \tag{27}$$

or explicitly by

$$H_n(x) = (2x)^n - \frac{n(n-1)}{1!}(2x)^{n-2} + \frac{n(n-1)(n-2)(n-3)}{2!}(2x)^{n-4} - + \ldots \tag{27'}$$

The first of these polynomials are

$$\begin{aligned} H_0(x) &= 1 \quad H_1(x) = 2x \\ H_2(x) &= 4x^2 - 2 \quad H_3(x) = 8x^3 - 12x \\ H_4(x) &= 16x^4 - 48x^2 + 12 \ldots \end{aligned} \tag{27''}$$

Considering next the proper values, we get from (25) and (23)

$$E_n = \frac{2n+1}{2} h\nu_0; \quad n = 0, 1, 2, 3, \ldots \tag{25'}$$

---

[16] Cf. Courant-Hilbert, *Methods of Mathematical Physics*, i. (Berlin, Springer, 1924), v. § 9, p. 261, eqn. 43, and further ii. § 10, 4, p. 76.

Thus as quantum levels appear so-called "half-integral" multiples of the "quantum of energy" peculiar to the oscillator, i.e. the *odd* multiples of $\frac{h\nu_0}{2}$. The intervals between the levels, which alone are important for the radiation, are the same as in the former theory. It is remarkable that our quantum levels are *exactly* those of Heisenberg's theory. In the theory of *specific heat* this deviation from the previous theory is not without significance. It becomes important first when the proper frequency ν_0 *varies* owing to the dissipation of heat. Formally it has to do with the old question of the "zero-point energy," which was raised in connection with the choice between the first and second forms of Planck's Theory. By the way, the additional term $\frac{h\nu_0}{2}$ also influences the law of the *band-edges*.

The *proper functions* (26) become, if we reintroduce the original q from (24) and (23),

$$\psi_n(q) = e^{-\frac{2\pi^2 \nu_0 q^2}{h}} H_n\left(2\pi q \sqrt{\frac{\nu_0}{h}}\right). \tag{26'}$$

Consideration of (27″) shows that the first function is a *Gaussian Error-curve*; the second vanishes at the origin and for x positive corresponds to a "Maxwell distribution of velocities" in two dimensions, and is continued in the manner of an odd function for x negative. The third function is even, is negative at the origin, and has two

symmetrical zeros at $\pm\frac{1}{\sqrt{2}}$, etc. The curves can easily be sketched roughly and it is seen that the roots of consecutive polynomials *separate* one another. From (26') it is also seen that the characteristic points of the proper functions, such as half-breadth (for $n = 0$), zeros, and maxima, are, as regards order of magnitude, within the range of the classical vibration of the oscillator. For the classical *amplitude* of the nth vibration is readily found to be given by

$$q_n = \frac{\sqrt{E_n}}{2\pi\nu_0} = \frac{1}{2\pi}\sqrt{\frac{h}{\nu_0}}\sqrt{\frac{2n+1}{2}}. \tag{28}$$

Yet there is in general, as far as I see, no definite meaning that can be attached to the *exact* abscissa of the classical *turning points* in the graph of the proper function. It may, however, be conjectured, because the turning points have *this* significance for the phase space wave, that, at them, the square of the velocity of propagation becomes *infinite* and at greater distances becomes *negative*. In the differential equation (22), however, this only means the *vanishing* of the coefficient of ψ and gives rise to no singularities.

I would not like to suppress the remark here (and it is valid quite generally, not merely for the oscillator), that nevertheless this vanishing and becoming imaginary of the velocity of propagation is something which is very characteristic. It is the analytical reason for the selection of definite proper values, merely through the condition that the function should remain finite. I would like to illustrate this further. A wave equation with a *real* velocity of propagation means just this: there is an *accelerated* increase in the value of the function at all those points where its value is *lower* than the average of the values at neighbouring points, and vice versa. Such an equation, if not immediately and lastingly as in case of the *equation for the conduction of heat*, yet in the course of time, causes a *levelling* of extreme values and does not permit at any point an excessive growth of the function. A wave equation with an *imaginary* velocity of propagation means the exact opposite: values of the function above the average of surrounding values experience an *accelerated increase* (or retarded decrease), and vice versa. We see, therefore, that a function represented by such an equation is in the greatest danger of growing beyond all bounds, and we must order matters skilfully to preserve it from this danger. The sharply defined proper values are just what makes this possible. Indeed, we can see in the example treated in Part I. that the demand for sharply defined proper values immediately ceases as soon as we choose the quantity E to be *positive*, as this makes the wave velocity real throughout all space.

After this digression, let us return to the oscillator and ask ourselves if anything is altered when we allow it two or more degrees of freedom (space oscillator, rigid body). If *different* mechanical proper frequencies (ν_0-values) belong to the separate coordinates, then nothing is changed. ψ is taken as the *product* of functions, each of a single coordinate, and the problem splits up into just as many separate problems of the type treated above as there are coordinates present. The proper functions are products of Hermite orthogonal functions, and the proper values of the whole problem appear as sums of those of the separate problems, taken in every possible combination. No proper value (for the whole system) is multiple, if we presume that there is no rational relation between the ν_0-values.

If, however, there is such a relation, then the same manner of treatment is still *possible*, but it is certainly not *unique*. Multiple proper values appear and the "separation" can certainly be effected in other coordinates, e.g. in the case of the isotropic space oscillator in spherical polars.[17]

The proper values that we get, however, are certainly in each case exactly the same, at least in so far as we are able to prove the "completeness" of a system of proper functions, obtained in *one* way. We recognise here a complete parallel to the well-known relations which the method of the previous quantisation meets with in the case of *degeneracy*. Only in one point there is a not unwelcome formal difference. If we applied the Sommerfeld-Epstein quantum conditions *without*

---

[17]We are led thus to an equation in r, which may be treated by the method shown in the Kepler problem of Part I. Moreover, the *one*-dimensional oscillator leads to the same equation if q^2 be taken as variable. I originally solved the problem directly in *that* way. For the hint that it was a question of Hermite polynomials, I have to thank Herr E. Fues. The polynomial appearing in the Kepler problem (eqn. 18 of Part I.) is the $(2n+1)$th differential coefficient of the $(n+l)$th polynomial of Laguerre, as I subsequently found.

regard to a possible degeneracy then we always got the same energy levels, but reached different conclusions as to the paths permitted, according to the choice of coordinates.

Now that is *not* the case here. Indeed we come to a completely different system of proper functions, if we, for example, treat the vibration problem corresponding to unperturbed Kepler motion in *parabolic* coordinates instead of the polars used in Part I. However, it is not just the *single proper vibration* that furnishes a *possible state of vibration*, but an arbitrary, finite or infinite, *linear aggregate* of such vibrations. And as such the proper functions found in any second way may always be represented; namely, they may be represented as linear aggregates of the proper functions found in an arbitrary way, provided the latter form a *complete* system.

The question of how the energy is really distributed among the proper vibrations, which has not been taken into account here up till now, will, of course, have to be faced some time. Relying on the former quantum theory, we will be disposed to assume that in the degenerate case only the energy of the set of vibrations belonging to one definite proper value must have a certain prescribed value, which in the non-degenerate case belongs to one single proper vibration. I would like to leave this question still *quite* open – and also the question whether the discovered "energy levels" are really energy steps of the *vibration process* or whether they *merely* have the significance of its frequency. If we accept the beat theory, then the meaning of energy levels is no longer necessary for the explanation of sharp emission frequencies.

2. *Rotator with Fixed Axis*

On account of the lack of potential energy and because of the *Euclidean* fine element, this is the simplest conceivable example of vibration theory. Let A be the moment of inertia and ϕ the angle of rotation, then we clearly obtain as the vibration equation

$$\frac{1}{A}\frac{d^2\psi}{d\phi^2} + \frac{8\pi^2 E}{h^2}\psi = 0, \tag{29}$$

which has the solution

$$\psi = \frac{\sin}{\cos}\left[\sqrt{\frac{8\pi^2 EA}{h^2}}\cdot\phi\right]. \tag{30}$$

Here the argument must be an *integral* multiple of ϕ, simply because otherwise ψ would neither be single-valued nor continuous throughout the range of the coordinate ϕ, as we know $\phi+2\pi$ has the same significance as ϕ. This condition gives the well-known result

$$E_n = \frac{n^2 h^2}{8\pi^2 A} \tag{31}$$

in *complete* agreement with the former quantisation.

   No meaning, however, can be attached to the result of the application to band spectra. For, as we shall learn in a moment, it is a peculiar fact that our theory gives *another* result for the rotator with *free* axis. *And this is true in general.* It is not allowable in the applications of wave mechanics, to think of the freedom of movement of the system as being more strictly limited, in order to simplify calculation, than it *actually* is, even when we know from the integrals of the mechanical equations that in a single movement certain definite freedoms are not made use of. For micro-mechanics, the fundamental system of mechanical equations is absolutely incompetent; the single paths with which it deals have now no separate existence. A wave process fills the *whole* of the phase space. It is well known that even the *number* of the dimensions in which a wave process takes place is very significant.

3. *Rigid Rotator with Free Axis*

If we introduce as coordinates the polar angles θ, ϕ of the radius from the nucleus, then for the kinetic energy as a function of the momenta we get

$$T = \frac{1}{2A}\left(p_\theta^2 + \frac{p_\phi^2}{\sin^2\theta}\right). \tag{32}$$

According to its form this is the kinetic energy of a particle constrained to move on a spherical surface. The Laplacian operator is thus simply that part of the spatial Laplacian operator which depends on the polar angles, and the vibration equation (18″) takes the following form,

$$\frac{1}{\sin\theta}\frac{\partial}{\partial\theta}\left(\sin\theta\frac{\partial\psi}{\partial\theta}\right) + \frac{1}{\sin^2\theta}\frac{\partial^2\psi}{\partial\phi^2} + \frac{8\pi^2 AE}{h^2}\psi = 0. \tag{33}$$

The postulation that ψ should be single-valued and continuous on the spherical surface leads to the proper value condition

$$\frac{8\pi^2 A}{h^2}E = n(n+1); \qquad n = 0, 1, 2, 3, \ldots \tag{34}$$

The proper functions are known to be spherical surface harmonics. The energy levels are, therefore,

$$E_n = \frac{n(n+1)h^2}{8\pi^2 A}; \qquad n = 0, 1, 2, 3, \ldots \tag{34'}$$

This definition is different from all previous statements (except perhaps that of Heisenberg ?). Yet, from various arguments from experiment we were led to put "half-integral" values for n in formula (31). It is easily seen that (34′) gives practically the same as (31) with half-integral values of n. For

$$n(n+1) = (n + \tfrac{1}{2})^2 - \tfrac{1}{4}.$$

The discrepancy consists only of a small additive constant; the level *differences* in (34) are the same as are got from "half-integral quantisation". This is true also for the application to short-wave bands, where the moment of inertia is not the same in the initial and final states, on account of the "electronic jump". For at most a small constant additional part comes in for all lines of a band, which is swamped in the large "electronic term" or in the "nuclear vibration term". Moreover, our previous analysis does not permit us to speak of this small part in any more definite way than as, say,

$$\frac{1}{4} \frac{h^2}{8\pi^2} \left(\frac{1}{A} - \frac{1}{A'} \right).$$

The notion of the moment of inertia being fixed by "quantum conditions" for electronic motions and nuclear vibrations follows naturally from the whole line of thought developed here. We will show in the next section how we can treat, approximately at least, the nuclear vibrations and the rotations of the diatomic molecule simultaneously by a synthesis of the cases[18] considered in 1 and 3.

I should like to mention also that the value $n = 0$ corresponds not to the vanishing of the wave function ψ but to a *constant* value for it, and accordingly to a vibration with amplitude constant over the whole sphere.

4. Non-rigid Rotator (Diatomic Molecule)

According to the observation at the end of section 2, we must state the problem initially with all the six degrees of freedom that the rotator really possesses. Choose Cartesian coordinates for the two molecules, viz. x_1, y_1, z_1; x_2, y_2, z_2, and let the masses be m_1 and m_2, and r be their distance apart. The potential energy is

$$V = 2\pi^2 \nu_0^2 \, \mu (r - r_0)^2, \tag{35}$$

---

[18]Cf. A. Sommerfeld, *Atombau und Spektrallinien*, 4th edit., p. 833. We do not consider here the additional non-harmonic terms in the potential energy.

where

$$r^2 = (x_1 - x_2)^2 + (y_1 - y_2)^2 + (z_1 - z_2)^2.$$

Here

$$\mu = \frac{m_1 m_2}{m_1 + m_2} \tag{36}$$

may be called the "resultant mass" Then v 0 is the mechanical proper frequency of the nuclear vibration, regarding the line joining the nuclei as fixed, and r_0 is the distance apart for which the potential energy is a minimum. These definitions are all in the sense of the usual mechanics.

For the vibration equation (18″) we get the following:

$$\begin{cases} \dfrac{1}{m_1}\left(\dfrac{\partial^2\psi}{\partial x_1^2} + \dfrac{\partial^2\psi}{\partial y_1^2} + \dfrac{\partial^2\psi}{\partial z_1^2}\right) + \dfrac{1}{m_2}\left(\dfrac{\partial^2\psi}{\partial x_2^2} + \dfrac{\partial^2\psi}{\partial y_2^2} + \dfrac{\partial^2\psi}{\partial z_2^2}\right) \\[2mm] + \dfrac{8\pi^2}{h^2}[E - 2\pi^2 v_0^2 \mu(r - r_0)^2]\psi = 0. \end{cases} \tag{37}$$

Introduce new independent variables $x, y, z, \xi, \eta, \zeta$, where

$$\begin{aligned} x &= x_1 - x_2; \ (m_1 + m_2)\xi = m_1 x_1 + m_2 x_2 \\ y &= y_1 - y_2; \ (m_1 + m_2)\eta = m_1 y_1 + m_2 y_2 \\ z &= z_1 - z_2; \ (m_1 + m_2)\zeta = m_1 z_1 + m_2 z_2. \end{aligned} \tag{38}$$

The substitution gives

$$\begin{cases} \dfrac{1}{\mu}\left(\dfrac{\partial^2\psi}{\partial x_1^2} + \dfrac{\partial^2\psi}{\partial y_1^2} + \dfrac{\partial^2\psi}{\partial z_1^2}\right) + \dfrac{1}{m_1 + m_2}\left(\dfrac{\partial^2\psi}{\partial \xi^2} + \dfrac{\partial^2\psi}{\partial \eta^2} + \dfrac{\partial^2\psi}{\partial \zeta^2}\right) \\[2mm] + [a'' - b'(r - r_0)^2]\psi = 0. \end{cases} \tag{37'}$$

where for brevity

$$a'' = \frac{8\pi^2 E}{h^2}, \qquad b' = \frac{16\pi^4 v_0^2 \mu}{h^2} \tag{39}$$

Now we can put for ψ the product of a function of the relative coordinates x, y, z, and a function of the coordinates of the centre of mass ξ, η, ζ:

$$\psi = f(x, y, z)\, g(\xi, \eta, \zeta). \tag{40}$$

For g we get the defining equation

$$\frac{1}{m_1 + m_2}\left(\frac{\partial^2 \psi}{\partial \xi^2} + \frac{\partial^2 \psi}{\partial \eta^2} + \frac{\partial^2 \psi}{\partial \zeta^2}\right) + \text{const.}\ g = 0. \tag{41}$$

This is of the same form as the equation for the motion, under no forces, of a particle of mass $m_1 + m_2$. The constant would in this case have the meaning

$$\text{const.} = \frac{8\pi^2 E_t}{h^2}, \tag{42}$$

where E_t is the energy of translation of the said particle. Imagine this value inserted in (41). The question as to the values of E_t admissible as proper values depends now on this, whether the whole infinite space is available for the original coordinates and hence for those of the centre of gravity without new potential energies coming in, or not. In the first case every non-negative value is permissible and every negative value not permissible. For when E_t is not negative and *only* then, (41) possesses solutions which do not vanish identically and yet remain finite in all space. If, however, the molecule is situated in a "vessel", then the latter must supply boundary conditions for the function g, or in other words, equation (41), on account of the introduction of further potential energies, will alter its form very abruptly at the walls of the vessel, and thus a discrete set of E_t-values will be selected as proper values. It is a question of the "Quantisation of the motion of translation," the main points of which I have lately discussed, showing that it leads to Einstein's Gas Theory.[19]

---

[19] *Physik . Ztschr.* 27, p. 95, 1926.

For the factor f of the vibration function ψ, depending on the relative coordinates x, y, z, we get the defining equation

$$\frac{1}{\mu} \left(\frac{\partial^2 f}{\partial x^2} + \frac{\partial^2 f}{\partial y^2} + \frac{\partial^2 f}{\partial z^2} \right) + [a' - b'(r - r_0)^2]f = 0, \quad (43)$$

where for brevity we put

$$a' = \frac{8\pi^2(E - E_t)}{h^2}. \quad (39')$$

We now introduce instead of x, y, z, the spherical polars r, θ, ϕ (which is in agreement with the previous use of r). After multiplying by μ we get

$$\frac{1}{r^2} \frac{\partial}{\partial r} \left(r^2 \frac{\partial f}{\partial r} \right) + \frac{1}{r^2} \left\{ \frac{1}{\sin\theta} \frac{\partial}{\partial\theta} \left(\sin\theta \frac{\partial f}{\partial\theta} \right) + \frac{1}{\sin^2\theta} \frac{\partial^2 f}{\partial\phi^2} \right\}$$
$$+ [\mu a' - \mu b'(r - r_0)^2]f = 0.$$
$$(43')$$

Now break up f. The factor depending on the angles is a surface harmonic. Let the order be n. The curled bracket is $-n(n + 1)f$. Imagine this inserted and for simplicity let f now stand for the factor depending on r. Then introduce as new *dependent* variable

$$\chi = rf, \quad (44)$$

and as new *independent* variable

$$\rho = r - r_0. \quad (45)$$

The substitution gives

$$\frac{\partial^2 \chi}{\partial\rho^2} + \left[\mu a' - \mu b'\rho^2 - \frac{n(n + 1)}{(r_0 + \rho)^2} \right] \chi = 0. \quad (46)$$

To this point the analysis has been exact. Now we will make an approximation, which I well know requires a stricter

64

justification than I will give here. Compare (46) with equation
(22′) treated earlier. They agree in form and only differ in the
coefficient of the unknown function by terms of the relative
order of magnitude of $\frac{\rho}{r_0}$. This is seen, if we develop thus:

$$\frac{n(n+1)}{(r_0+\rho)^2} = \frac{n(n+1)}{r_0^2}\left(1 - \frac{2\rho}{r_0} + \frac{3\rho^2}{r_0^2} - + \ldots\right), \qquad (47)$$

substitute in (46), and arrange in powers of ρ/r_0. If we introduce for ρ a new variable differing only by a small constant,
viz.

$$\rho' = \rho - \frac{n(n+1)}{r_0^3\left(\mu b' + \frac{3n(n+1)}{r_0^4}\right)} \qquad (48)$$

then equation (46) takes the form

$$\frac{\partial^2\chi}{\partial\rho'^2} + \left(a - b\rho'^2 + \left[\frac{\rho'}{r_0}\right]\right)\chi = 0, \qquad (46')$$

where we have put

$$\begin{cases} a = \mu a' - \dfrac{n(n+1)}{r_0^2}\left(1 - \dfrac{n(n+1)}{r_0^4\mu b' + 3n(n+1)}\right) \\ b = \mu b' + \dfrac{3n(n+1)}{r_0^4}. \end{cases} \qquad (49)$$

The symbol $\left[\frac{\rho'}{r_0}\right]$ in (46′) represents terms which are small
compared with the retained term of the order of $\frac{\rho'}{r_0}$.

Now we know that the *first* proper functions of equation
(22′), to which we now compare (46′), only differ markedly
from zero in a small range on both sides of the origin. Only
those of higher order stretch gradually further out. For mod-
erate orders, the domain for equation (46′), if we *neglect* the
term $\left[\frac{\rho'}{r_0}\right]$ and bear in mind the order of magnitude of molec-
ular constants, is indeed small compared with r_0. We thus
conclude (without rigorous proof, I repeat), that we can in

this way obtain a useful approximation for the first proper functions, within the region where they differ at all markedly from zero, and also for the first proper values. From the proper value condition (25) and omitting the abbreviations (49), (39′), and (39), though introducing the small quantity

$$\epsilon = \frac{n(n+1)h^2}{16\pi^4\nu_0^2\mu^2 r_0^4} = \frac{n(n+1)h^2}{16\pi^4\nu_0^2 A^2} \tag{50}$$

instead, we can easily derive the following *energy steps*,

$$\begin{cases} E = E_t + \dfrac{n(n+1)h^2}{8\pi^2 A}\left(1 - \dfrac{\epsilon}{1+3\epsilon}\right) + \dfrac{2l+1}{2}h\nu_0\sqrt{1+3\epsilon} \\ (n = 0,1,2,\ldots; \qquad l = 0,1,2,\ldots), \end{cases} \tag{51}$$

where

$$A = \mu r_0^2 \tag{52}$$

is still written for the *moment of inertia*.

In the language of classical mechanics, ϵ is the square of the ratio of the frequency of rotation to the vibration frequency ν_0; it is therefore really a small quantity in the application to the molecule, and formula (51) has the usual structure, apart from this small correction and the other differences already mentioned. It is the synthesis of (25′) and (34′) to which E_t is added as representing the energy of translation. It must be emphasized that the value of the approximation is to be judged not only by the smallness of ϵ but also by l not being too large. *Practically*, however, only small numbers have to be considered for l.

The ϵ-corrections in (51) do *not yet* take account of deviations of the nuclear vibrations from the pure harmonic type. Thus a comparison with Kratzer's formula (*vide* Sommerfeld, *loc. cit.*) and with experience is impossible. I only desired to mention the case provisionally, as an example showing that the intuitive idea of the *equilibrium configuration* of the nuclear system retains its meaning in undulatory mechanics also,

66

and showing the manner in which it does so, provided that the wave amplitude ψ is different from zero practically only in a small neighbourhood of the equilibrium configuration. The direct interpretation of this wave function of *six* variables in *three*-dimensional space meets, at any rate initially, with difficulties of an abstract nature.

The rotation- vibration-problem of the diatomic molecule will have to be re-attacked presently, the non-harmonic terms in the energy of binding *being taken into account.* The method, selected skilfully by Kratzer for the classical mechanical treatment, is also suitable for undulatory mechanics. If, however, we are going to push the calculation as far as is necessary for the fineness of band structure, then we must make use of the theory of the *perturbation of proper values and functions*, that is, of the alteration experienced by a definite proper value and the appertaining proper functions of a differential equation, when there is added to the coefficient of the unknown function in the equation a small "disturbing term." This "perturbation theory" is the complete counterpart of that of classical mechanics, except that it is simpler because in undulatory mechanics we are always in the domain of *linear* relations. As a first approximation we have the statement that the perturbation of the proper value is equal to the perturbing term averaged "overr the undisturbed motion."

The perturbation theory broadens the analytical range of the new theory extraordinarily. As an important practical success, let me say here that the *Stark effect* of the first order will be found to be really completely in accord with Epstein' formula, which has become unimpeachable through the confirmation of experience.

Zürich, Physical Institute of the University.
(Received February 23, 1926.)

THE CONTINUOUS TRANSITION FROM MICRO- TO MACRO-MECHANICS

(*Die Naturwissenschaften*, 28, pp. 664-666, 1926)

Building on ideas of de Broglie[1] and Einstein,[2] I have tried to show[3] that the usual differential equations of mechanics, which attempt to define the coordinates of a mechanical system as functions of the time, are no longer applicable for "small" systems; instead there must be introduced a certain partial differential equation, which defines a variable ψ ("wave function") as a function of the coordinates and the time. As in the differential equation of a vibrating string or of any other vibrating system, ψ is given as a superposition of pure time harmonic (i.e. "sinusoidal") vibrations, the frequencies of which agree exactly with the spectroscopic "term frequencies" of the micro-mechanical system. For example, in the case of

---

[1] L. de Broglie, *Ann. de Physique* (10), 3, p. 22, 1925 (Thèses, Paris, 1924).

[2] A. Einstein, *Berlin Ber.* 1925, p. 9 *et seq.*

[3] *Ann. d. Physik*; the essays here collected.

68

the linear Planck oscillator[4] where the energy function is

$$\frac{m}{2}\left(\frac{dq}{dt}\right)^2 + 2\pi^2\nu_0^2 mq^2, \qquad (1)$$

when we put, instead of the displacement q, the dimensionless variable

$$x = q \cdot 2\pi\sqrt{\frac{m\nu_0}{h}}, \qquad (2)$$

we get ψ as the superposition of the following proper vibrations:[5]

$$\begin{cases} \psi_n = e^{-\frac{x^2}{2}} H_n(x)e^{2\pi i\nu_n t} \\ \left(\nu_n = \frac{2n+1}{2}\nu_0; \qquad n = 0,1,2,3,\ldots\right). \end{cases} \qquad (3)$$

The H_n's are the polynomials[6] named after Hermite. If they are multiplied by $e^{-\frac{x^2}{2}}$ and the "normalising factor" $(2^n\,n!)^{-\frac{1}{2}}$ they are called Hermite's orthogonal functions. They represent therefore the amplitudes of the proper vibrations.

The first five are represented in Fig. 1. The similarity between this and the well-known picture of the vibrations of a string is very great.

At first sight it appears very strange to try to describe a process, which we previously regarded as belonging to particle mechanics, by a system of such proper vibrations. For this chosen simple case, I would like to demonstrate here *in concreto* the transition to macroscopic mechanics, by showing that a *group* of proper vibrations of *high* order-number

---

[4]i.e. a particle of mass m which, moving in a straight line, is attracted towards a fixed point in it, with a force proportional to its displacement q from this point; according to the usual mechanics, such a particle executes sine vibrations of frequency ν_0.

[5]i means $\sqrt{-1}$. On the right-hand side the real part is to be taken, as usual.

[6]Cf. Courant-Hilbert, *Methoden der mathematischen Physik*, I. chap. ii. § 10, 4, p. 76 (Berlin, Springer, 1924).

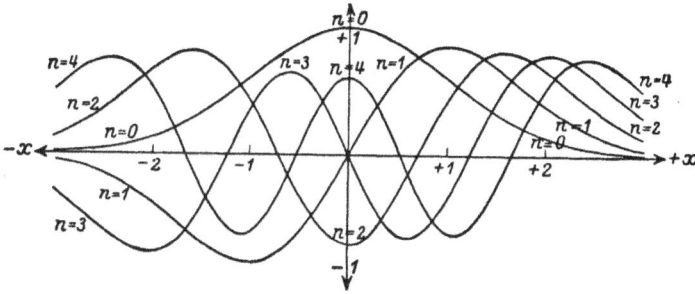

Fig. 1. The first five proper vibrations of the Planck oscillator according to undulatory mechanics. Outside of the region $-3 \leq x \leq +3$ represented here, all five functions approach the x-axis in monotonic fashion.

n ("quantum number") and of relatively small order-number differences ("quantum number differences") may represent a "particle," which is executing the "motion," expected from the usual mechanics, i.e. oscillating with the frequency ν_0. I choose a number $A \gg 1$ (i.e. great compared with 1) and form the following aggregate of proper vibrations:

$$
\begin{aligned}
\psi &= \sum_{n=0}^{\infty} \left(\frac{A}{2}\right)^n \frac{\psi_n}{n!} \\
&= e^{\pi i \nu_0 t} \sum_{n=0}^{\infty} \left(\frac{A}{2} e^{2\pi i \nu_0 t}\right)^n \frac{1}{n!} e^{-\frac{x^2}{2}} H_n(x).
\end{aligned}
\tag{4}
$$

Thus the *normalised* proper vibrations (see above) are taken with the coefficients

$$
\frac{A^n}{\sqrt{2^n n!}},
\tag{5}
$$

and this, as is easily seen,[7] results in the singling out of a relatively small group in the neighbourhood of the n-value

---

[7] $z^n/n!$ has, *as function of n, for large values of z*, a single extremely high and relatively very sharp maximum at $n = z$. By taking square roots and with $z = A^2/2$, we get the series of numbers (5).

given by

$$n = \frac{A^2}{2}. \qquad (6)$$

The summation of the series (4) is made possible by the following identity[8] in x and s:

$$\sum_{n=0}^{\infty} \frac{s^n}{n!} e^{-\frac{x^2}{2}} H_n(x) = e^{-s^2+2sx-\frac{x^2}{2}}. \qquad (7)$$

Thus

$$\psi = e^{\pi i \nu_0 t - \frac{A^2}{4} e^{4\pi i \nu_0 t} + Ax e^{2\pi i \nu_0 t} - \frac{x^2}{2}}. \qquad (8)$$

Now we take, as is provided for, the real part of the right-hand side and after a short calculation obtain

$$\psi = e^{\frac{A^2}{4} - \frac{1}{2}(x - A\cos 2\pi\nu_0 t)^2} \times$$
$$\times \cos\left[\pi\nu_0 t + (A\sin 2\pi\nu_0 t)\cdot\left(x - \frac{A}{2}\cos 2\pi\nu_0 t\right)\right]. \qquad (9)$$

This is the *final result,* in which the first factor is our first interest. It represents a relatively tall and narrow "hump," of the form of a "Gaussian error-curve," which at a given moment lies in the neighbourhood of the position

$$x = A\sin 2\pi\nu_0 t. \qquad (10)$$

The breadth of the hump is of the order of magnitude unity and therefore very small compared with A, by hypothesis. According to (10), the hump oscillates under exactly the same law as would operate in the usual mechanics for a particle having (1) as its energy function. The amplitude in terms of x is A, and thus in terms of q is

$$a = \frac{A}{2\pi}\sqrt{\frac{h}{m\nu_0}}. \qquad (11)$$

---

[8]Courant-Hilbert, *loc. cit.* eqn. (58).

Ordinary mechanics gives for the *energy* of a particle of mass m, which oscillates with this amplitude and with frequency ν_0,

$$2\pi^2 a^2 \nu_0^2 m = \frac{A^2}{2} h\nu_0, \tag{12}$$

i.e. from (6) exactly $nh\nu_0$, where n is the average quantum number of the selected group. The "correspondence" is thus complete in this respect also.

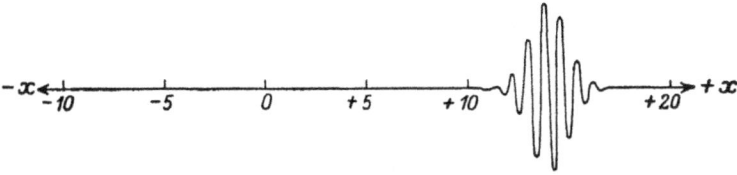

Fig. 2. Oscillating wave group as the representation of a particle in wave mechanics.

The *second* factor in (9) is in general a function whose absolute value is small compared with unity, and which varies very rapidly with x and also t. It ploughs many deep and narrow furrows in the profile of the first factor, and makes a *wave group* out of it, which is represented – schematically only – in Fig. 2. The x-scale of Fig. 2 is naturally much smaller than that of Fig. 1; Fig. 2 requires to be magnified five times before being directly compared with Fig. 1. A more exact consideration of the second factor of (9) discloses the following interesting details, which cannot be seen in Fig. 2, which only represents *one* stage. The *number* and *breadth* of the "furrows" or "wavelets" within the particle vary with the time. The wavelets are most *numerous* and *narrowest* when passing through the centre $x = 0$; they become *completely smoothed out* at the turning points $x = \pm A$, because there, by (10), $\cos 2\pi\nu_0 t = \pm 1$ and thus $\sin 2\pi\nu_0 t$, becomes equal to zero, so that the second factor of (9) is absolutely independent of x. The entire extension of the wave group ("density of

72

the particle") remains, however, always the same. The variability of the "corrugation" is to be conceived as depending on the *velocity*, and, as such, is completely intelligible from all general aspects of undulatory mechanics – but I do not wish to discuss this further at present.

Our wave group always remains compact, and does *not* spread out into larger regions as time goes on, as we were accustomed to make it do, for example, in optics. It is admitted that this does not mean much in one dimension, and that a hump on a string will behave quite similarly. But it is easily seen that, by multiplying together two or three expressions like (4), written in x, in y , and in z respectively, we can represent also the *plane* and the *spatial* oscillator respectively, i.e. a plane or spatial wave group which moves round a harmonic ellipse.[9] Also such a wave group will remain compact, in contrast, e.g., to a wave packet in classical optics, which is dissipated in the course of time. The distinction may originate in the fact that our group is built up out of separate *discrete* harmonic components, and not out of a *continuum* of such.

I wish to mention, in conclusion, that a general additive constant, C, let us say, which should be added to all the ν_n's in (3), (and corresponds to the "rest-energy" of the particle) does not alter the essentials. It only affects the square bracket in (9), adding $2\pi Ct$ thereto. Hence the oscillations *within* the wave group become very much quicker with respect to the *time*, while the oscillation of the group as a whole, given by (10), and its "corrugation", remain quite unaffected.

We can definitely foresee that, in a similar way, wave groups can be constructed which move round highly quantised

---

[9]We may point out, in passing, the interesting fact that the quantum levels of the *plane* oscillator are *integral*, but for the *spatial* oscillator they again become "half-integral." Similarly for the rotator. This half-integralness, which is spectroscopically so significant, is thus connected with the "oddness" of the number of the dimensions of space.

Kepler ellipses and are the representation by wave mechanics of the hydrogen electron. But the technical difficulties in the calculation are greater than in the especially simple case which we have treated here.

74

On the Relation between the Quantum Mechanics of Heisenberg, Born, and Jordan, and that of Schrödinger

(*Annalen der Physik* (4), vol. 79, 1926)

§ 1. Introduction and Abstract

Considering the extraordinary differences between the starting-points and the concepts of Heisenbergs quantum mechanics[1] and of the theory which has been designated "undulatory" or "physical" mechanics,[2] and has lately been described here, it is very strange that these two new theories agree *with one another* with regard to the known facts, where they differ from the old quantum theory. I refer, in particular, to the peculiar "half-integralness" which arises in connection with the oscillator and the rotator. That is really very remarkable, because starting-points, presentations, methods, and in fact the

---

[1] W. Heisenberg, *Ztschr.f. Phys.* 33, p. 879, 1925; M. Born and P. Jordan, *idem* 34, p. 858, 1925, and 35, p. 557, 1926 (the latter in collaboration with Heisenberg). I may be allowed, for brevity's sake, to replace the three names simply by Heisenberg, and to quote the last two essays as "Quantum Mechanics I. and II." Interesting contributions to the theory have also been made by P. Dirac, *Proc. Roy. Soc.*, London, 109, p. 642. 1925, and *idem* 110, p. 561, 1926.

[2] E. Schrodinger. Parts I. and II. in this collection. These parts will be continued quite independently of the present paper, which is only intended to serve as a connecting link.

whole mathematical apparatus, seem fundamentally different. Above all, however, the departure from classical mechanics in the two theories seems to occur in diametrically opposed directions. In Heisenberg's work the classical continuous variables are replaced by systems of discrete numerical quantities (matrices), which depend on a pair of integral indices, and are defined by *algebraic* equations. The authors themselves describe the theory as a "true theory of a discontinuum."[3] On the other hand, wave mechanics shows just the reverse tendency; it is a step from classical point-mechanics towards a *continuum-theory.* In place of a process described in terms of a finite number of dependent variables occurring in a finite number of total differential equations, we have a continuous *field-like* process in configuration space, which is governed by a single *partial* differential equation, derived from a principle of action. This principle and this differential equation replace the equations of motion *and* the quantum conditions of the older "classical quantum theory."[4]

In what follows the very intimate *inner connection* between Heisenberg's quantum mechanics and my wave mechanics will be disclosed. From the formal mathematical standpoint, one might well speak of the *identity* of the two theories. The train of thought in the proof is as follows.

Heisenberg's theory connects the solution of a problem in quantum mechanics with the solution of a system of an infinite number of algebraic equations, in which the unknowns – infinite matrices – are allied to the classical position- and momentum-coordinates of the mechanical system, and func-

---

[3]"Quantum Mechanics I." p. 879.

[4]My theory was inspired by L. de Broglie, *Ann. de Physique* (10) 3, p. 22, 1925 (Thèses, Paris, 1924), and by brief, yet infinitely far-seeing remarks of A. Einstein, *Berl. Ber.*, 1925, p. 9 *et seq.* I did not at all suspect any relation to Heisenberg's theory at the beginning. I naturally knew about his theory, but was discouraged, if not repelled, by what appeared to me as very difficult methods of transcendental algebra, and by the want of perspicuity (*Anschaulichkeit*).

tions of these, and obey peculiar *calculating rules*. (The relation is this: to *one* position-, *one* momentum-coordinate, or to *one* function of these corresponds always *one* infinite matrix.)

I will first show (§§ 2 and 3) how to each function of the position- and momentum-coordinates there may be related a matrix in such a manner, that these matrices, in *every case*, *satisfy* the formal calculating rules of Born and Heisenberg (among which I also reckon the so-called "quantum condition" or "interchange rule;" see below). This relation of matrices to functions is *general*; it takes no account of the *special* mechanical system considered, but is the same for all mechanical systems. (In other words: the particular Hamilton function does not enter into the connecting law.) However, the relation is still *indefinite* to a great extent. It arises, namely, from the *auxiliary introduction* of an *arbitrary* complete orthogonal system of functions having for domain *entire configuration space* (N.B. – *not* "pq-space," but "q-space"). The provisional *indefiniteness* of the relation lies in the fact that we can assign the *auxiliary* role to an *arbitrary* orthogonal system.

After matrices are thus constructed in a very general way, so as to satisfy the general rules, I will show the following in § 4. The *special* system of algebraic equations, which, in a *special* case, connects the *matrices* of the position and impulse coordinates with the *matrix* of the Hamilton function, and which the authors call "equations of motion," will be completely solved by assigning the auxiliary role to a *definite* orthogonal system, namely, to the system of *proper functions* of that partial differential equation which forms the basis of my wave mechanics. The solution of the natural *boundary-value problem* of this differential equation is *completely equivalent* to the solution of Heisenberg's algebraic problem. *All* Heisenberg's matrix elements, which may interest us from the surmise that they define "transition probabilities" or "line intensities", can be actually evaluated *by differentiation and quadra-*

ture, as soon as the *boundary-value problem* is solved. More-over, in wave mechanics, these matrix elements, or quantities that are closely related to them, have the perfectly clear sig-nificance of amplitudes of the partial oscillations of the atom's electric moment. The intensity and polarisation of the emit-ted light is thus intelligible *on the basis of the Maxwell-Lorentz theory.* A short preliminary sketch of this relationship is given in § 5.

§ 2. The Co-ordination of an Operator and of a Matrix with a Well-arranged Function-symbol and the Establishment of the Product Rule

The starting-point in the construction of matrices is given by the simple observation that Heisenberg's peculiar calcu-lating laws for functions of the *double* set of n quantities, $q_1, q_2, \ldots q_n; p_1, p_2, \ldots p_n$ (position- and canonically conjugate momentum-coordinates) agree exactly with the rules, which *ordinary analysis* makes *linear differential operators* obey in the domain of the *single* set of n variables, $q_1, q_2, \ldots q_n$. So the coordination has to occur in such a manner that each p_l in the *function* is to be replaced by the operator $\frac{\partial}{\partial q_l}$. Actu-ally the operator $\frac{\partial}{\partial q_l}$ is exchangeable with $\frac{\partial}{\partial q_m}$, where m is arbitrary, but with q_m only, if $m \neq l$. The operator, obtained by interchange and subtraction when $m = l$, viz.

$$\frac{\partial}{\partial q_l} q_l - q_l \frac{\partial}{\partial q_l},\qquad (1)$$

when applied to any arbitrary function of the q's, *reproduces* the function, i.e. this operator gives *identity*. This simple fact will be reflected in the domain of matrices as Heisenberg's interchange rule.

After this preliminary survey, we turn to systematic con-struction. Since, as noticed above, the interchangeability does not *always* hold good, then a definite operator does not cor-

respond uniquely to a definite "function in the usual sense" of
the q's and p's, but to a "function-symbol written in a definite
way". Moreover, since we can perform only the operations of
addition and multiplication with the operators $\frac{\partial}{\partial q_k}$, the function of the q's and p's must be written as a regular power
series in p at least, before we substitute $\frac{\partial}{\partial q_l}$ for p_l. It is sufficient to carry out the process for a single term of such a power
series, and thus for a function of the following construction:

$$F(q_k, p_k) =$$
$$= f(q_1 \ldots q_n) p_r p_s p_t \; g(q_1 \ldots q_n) p_{r'} \; h(q_1 \ldots q_n) p_{r''} p_{s''} \ldots \tag{2}$$

We wish to express this as a "well-arranged[5] function-symbol"
and relate it to the following operator,

$$[F, \bullet] = f(q_1 \ldots q_n) K^3 \frac{\partial^3}{\partial q_r \partial q_s \partial q_t} \; g(q_1 \ldots q_n) K \frac{\partial}{\partial q_{r'}} \times$$
$$\times \; h(q_1 \ldots q_n) K^2 \frac{\partial^2}{\partial q_{r''} \partial q_{s''}} \ldots , \tag{3}$$

wherein, somewhat more generally than in the preliminary
survey, p_r is not replaced by $\frac{\partial}{\partial q_r}$ simply, but by $K \frac{\partial}{\partial q_r}$ and
K stands for a universal constant. As an abbreviation for
the operator arising out of the well-arranged function F, I
have introduced the symbol $[F, \bullet]$ in passing (i.e. only for
the purpose of the present proof). The function (in the usual
sense) of $q_1 \ldots q_n$ which is obtained by using the operator
on another function (in the usual sense), $u(q_1 \ldots q_n)$, will be
denoted by $[F, u]$. If G is another well-arranged function, then
$[G F, u]$ will denote the function u after the operator of F has
first been used on it, and *then* the operator of G; or, what is
defined to be the same, when the operator of $G F$ has been
used. Of course this is not generally the same as $[F G, u]$.

Now we connect a *matrix* with a well-arranged function,
like F, by means of its operator (3) and of an arbitrary complete orthogonal system having for its domain the whole of

---

[5] Or "well-ordered."

q-space. It is done as follows. For brevity we will simply write x for the group of variables $q_1, q_2, \ldots q_n$, as is usual in the theory of Integral Equations, and write $\int dx$ for an integral extending over the whole of q-space. The functions

$$u_1(x)\sqrt{\rho(x)}, \quad u_2(x)\sqrt{\rho(x)}, \quad u_3(x)\sqrt{\rho(x)}, \ldots \text{ad inf.} \quad (4)$$

are now to form a complete orthogonal system, normalised to 1. Let, therefore, in every case

$$\begin{cases} \int\int \rho(x)u_i(x)u_k(x)dx & = 0 \quad \text{for } i \neq k \\ & = 1 \quad \text{for } i = k. \end{cases} \quad (5)$$

Further, it is postulated that these functions vanish at the natural *boundary* of q-space (in general, infinity) in a way sufficient to cause the vanishing of certain boundary integrals which come in later on as secondary products after certain integrations by parts.

By the operator (3) we now relate the following *matrix*,

$$F^{kl} = \int \rho(x)u_k(x)[F, \, u_l(x)]dx, \quad (6)$$

to the function F represented by (2). (The way of writing the indices on the left-hand side must not suggest the idea of "contravariance; " from this point of view, here discarded, *one* index was formerly written above, and the other below; we write the matrix indices *above*, because later we will also have to write matrix elements, corresponding to the q's and p's, where the lower place is already occupied.) In words: a matrix element is computed by *multiplying* the function of the orthogonal system denoted by the *row*-index (whereby we understand always u_i, not $u_i\sqrt{\rho}$, by the "density function"' ρ, and by the result arising from using our operator on the orthogonal function corresponding to the *column*-index, and then by *integrating* the whole over the domain.[6]

It is not very difficult to show that additive and multiplicative combination of well-arranged functions or of the appertaining operators works out as matrix addition and matrix multiplication of the allied matrices. For addition the proof is trivial. For multiplication the proof runs as follows. Let G be any other well-arranged function, like F, and

$$G^{lm} = \int \rho(x)u_l(x)[G, u_m(x)]dx, \tag{7}$$

the matrix corresponding. We wish to form the product matrix

$$(FG)^{km} = \sum_l F^{kl}G^{lm}.$$

Before writing it, let us transform the expression (6) for F^{kl} as follows. By a series of integrations by parts, the operator $[F, \bullet]$ is "revolved" from the function $u_l(x)$ to the function $rho(x)u_k(x)$. By the expression "revolve" (instead of, say, "push") I wish to convey that the *sequence* of the operations reverses itself exactly thereby. The boundary integrals, which come in as "by-products," are to disappear (see above). The "revolved" operator, including the change of sign that accompanies an odd number of differentiations, will be denoted by $[\overline{F}, \bullet]$. For example, from (3) comes

$$[\overline{F}, \bullet] = (-1)^\tau \ldots K^2 \frac{\partial^2}{\partial q_{s''}\partial q_{r''}} h(q_1 \ldots q_n) K \frac{\partial}{\partial q_{r'}} \times$$
$$\times g(q_1 \ldots q_n) K^3 \frac{\partial^3}{\partial q_t \partial q_s \partial q_r} f(q_1 \ldots q_n), \tag{3'}$$

where τ = number of differentiations. By applying this sym-

---

[6]More briefly: F^{kl} is the kth "development coefficient" of the operator used on the function u_l.

bol, we have

$$F^{kl} = \int u_l(x)[\overline{F}, \rho(x)u_k(x)]dx. \qquad (6')$$

If we now calculate the product matrix, we get

$$\sum_l F^{kl}G^{lm}$$

$$= \sum_l \left\{ \int u_l(x)[\overline{F}, \rho(x)u_k(x)]dx \cdot \int \rho(x)u_l(x)[G, u_m(x)]dx \right\}$$

$$= \int [\overline{F}, \rho(x)u_k(x)][G, u_m(x)]dx.$$

$$(8)$$

The last equation is simply the so-called "relation of completeness" of our orthogonal system,[7] applied to the "development coefficients" of the functions

$$[G, u_m(x)] \quad \text{and} \quad \frac{1}{\rho(x)}[\overline{F}, \rho(x)u_k(x)].$$

Now in (8), let us revolve, by further integrations by parts, the operator $[F, \bullet]$ from the function $\rho(x)u_k(x)$ back again to the function $[G, u_m(x)]$ so that the operator regains its original form. We clearly get

$$(FG)^{km} = \sum_l F^{kl}G^{lm} = \int \rho(x)u_k(x)[FG, u_m(x)]dx. \quad (9)$$

On the left is the (km)th element of the product matrix, and on the right, by the law of connection (6), stands the (km)th element of the matrix, corresponding to the well-arranged product FG. Q.E.D.

---

[7]See, e.g., Courant-Hilbert, *Methods of Mathematical Physics*, I., p. 36. It is important to remember that the "relation of completeness" for the "development coefficients" is valid in every case, even when the developments themselves do *not* converge. If these do converge, then the equivalence (8) is directly evident.

§ 3. Heisenberg's Quantum Condition and the Rules for Partial Differentiation

Since operation (1) gave identity, then corresponding to the well-arranged function

$$p_l q_l - q_l p_l \tag{10}$$

we have the *operator*, multiplication by K, in accordance with our law of connection, in which we incorporated a universal constant K. Hence to function (10) corresponds the *matrix*

$$(p_l q_l - q_l p_l)^{ik} = K \int \rho(x) u_k(x) dx = 0 \text{ for } i \neq k \tag{11}$$
$$= K \text{ for } i = k.$$

That is Heisenberg's "quantum relation" if we put

$$K = \frac{h}{2\pi\sqrt{-1}}, \tag{12}$$

and this may be assumed to hold from now on. It is understood that we could have also found relation (11) by taking the two matrices allied to q_l and p_l, viz.

$$q_l^{ik} = \int q_l \rho(x) u_i(x) u_k dx,$$
$$p_l^{ik} = K \int \rho(x) u_i(x) \frac{\partial u_k(x)}{\partial q_l} dx, \tag{13}$$

multiplying them together in different sequence and subtracting the two results.

Let us now turn to the "rules for partial differentiation." A well-arranged function, like (2), is said to be differentiated partially with respect to $q - L$, when it is differentiated with respect to q_l without altering the succession of the factors *at each place* where q_l appears in it, and all these results are added.[8] Then it is easy to show that the following equation between the operators is valid:

$$\left[\frac{\partial F}{\partial q_l}, \bullet\right] = \frac{1}{K} \left[p_l F - F p_l, \bullet\right]. \tag{14}$$

The line of thought is this. Instead of really differentiating with respect to q_l, it is very convenient simply to prefix $p - L$ to the function; as it is, p_l must finally be replaced by $K\frac{\partial}{\partial q_l}$. Obviously I have to divide by K. Furthermore, when we apply the entire operator to any function u, the operator $\frac{\partial}{\partial q_l}$ will act not only on that part of F which contains q_l (as it *ought*), but also *wrongly* on the function u, affected by the entire operator. *This mistake is exactly corrected* by subtracting again the operation $[F p_l, \bullet]$!

Consider now partial differentiation with respect to a p_l. Its meaning for a well-arranged function, like (2), is a little simpler than in the case of $\frac{\partial}{\partial q_l}$, because the p's only appear as power products. We imagine every power of p_l to be resolved into single factors, e.g. think of $p_l p_l p_l$ instead of p_l^3, and we can then say: in partial differentiation with respect to p_l, every *separate* p_l that appears in F is to be *dropped* once, all the other p_l's remaining; all the results obtained are to be added. What will be the effect on the operator (3)? "Every separate $K\frac{\partial}{\partial q_l}$ is to be dropped once, and all the results so obtained are to be added."

I maintain that on this reasoning the operational equation

$$\left[\frac{\partial F}{\partial p_l}, \bullet\right] = \frac{1}{K} \left[F q_l - q_l F, \bullet\right] \tag{15}$$

is valid. Actually, I picture the operator $[F q_l, \bullet]$ as formed and now attempt to "push qi through F from right to left",

---

[8]We are naturally following Heisenberg faithfully in all these definitions. From a strictly logical standpoint the following proof is evidently superfluous, and we could have written down rules (14) and (15) right away, as they are proved in Heisenberg, and only depend upon the sum and product rules and the exchange rule (11) which we have proved.

that means, attempt to arrive at the operator $[q_l F, \bullet]$ through successive exchanges. This pushing through meets an obstacle only as often as I come against a $\frac{\partial}{\partial q_l}$. With the latter I may not interchange q_l simply, but have to replace

$$\frac{\partial}{\partial q_l} \quad \text{by} \quad 1 + q_l \frac{\partial}{\partial q_l} \tag{16}$$

in the interior of the operator. The secondary products of the interchange, which are yielded by this "uniformising", form just the desired "partial differential coefficients", as is easily seen. After the pushing-through process is finished, the operator $[q_l F, \bullet]$ still remains left over. It would be superfluous and therefore is explicitly subtracted in (15). Hence (15) is proved. The equations (14) and (15), which have been proved for *operators*, naturally hold good unchanged for the matrices belonging to the right-hand and left-hand sides, because by (6) one matrix, and one only, belongs to one linear operator (after the system $u_i(x)$ has been chosen once for all).[9]

---

[9]In passing it may bo noted that the converse of this theorem is also true, at least in the sense that certainly *not more* than *one* linear differential operator can belong to a given *matrix*, according to our connecting law (6), when the orthogonal system and the density function are prescribed. For in (6), let the F^{kl}'s be given, let $[F,]$ be the linear operator we are *seeking* and which we *presume* to *exist* y and let $\phi(x)$ be a function of $q_1, q_2, \ldots, q_n$, which is sectionally continuous and differentiable as often as necessary, but otherwise arbitrary. Then the *relation of completeness* applied to the functions $\phi(x)$ and $[F, u_k(x)]$ yields the following:

$$\int \rho(x)\phi(x)[F, u_k(x)]dx = \sum_l \int \rho(x)\phi(x)u_l(x)dx \cdot \int \rho(x)u_l(x)[F, u_k(x)]dx$$

The right-hand side can be regarded as definitely known, for in it occur only development coefficients of $\phi(x)$ and the prescribed matrix elements F^{lk}. By "revolving" (see above), we can change the left-hand side into the kth development coefficient of the function

$$\frac{[F, \rho(x)\phi(x)]}{\rho(x)}.$$

§ 4. The Solution of Heisenberg's Equations of Motion

We have now shown that matrices, constructed according to definitions (3) and (6) from well-arranged functions by the agency of an arbitrary, complete orthogonal system (4), satisfy all Heisenberg's calculating rules, including the interchange rule (11). Now let us consider a special mechanical problem, characterised by a definite Hamilton function

$$H(q_k, p_k). \tag{17}$$

The authors of quantum mechanics take this function over from *ordinary* mechanics, which naturally does *not* give it in a "well-arranged" form; for in ordinary analysis no stress is laid on the sequence of the factors. They therefore "normalise" or "symmetricalise" the function in a definite manner for their purposes. For example, the usual mechanical function $q_k p_k^2$ is replaced by

$$\frac{1}{2} \left(p_k^2 q_k + q_k p_k^2 \right)$$

or by

$$p_k q_k p_k$$

or by

$$\frac{1}{3} \left(p_k^2 q_k + p_k q_k p_k + q_k p_k^2 \right),$$

Thus all the development coefficients of this function are uniquely fixed, and thus so is the function itself (Courant-Hilbert, p. 37). Since, however, $\rho(x)$ was fixed before-hand and $\phi(x)$ is a quite arbitrary function, we can say: the result of the action of the *revolved* operator on an *arbitrary* function, provided, of course, it can be submitted to the operator at all, is fixed *uniquely* by the matrix F^{kl}. This can only mean that *the revolved operator* is uniquely fixed, for the notion of "operator" is logically identical with the whole of the results of its action. By revolving the revolved operator, we obtain uniquely the operator we have sought, itself.

It is to be noted that the *developability* of the functions which appear is *not* necessarily postulated – we have not proved that a linear operator, corresponding to an arbitrary matrix, *always exists*.

which are all the same, according to (11). This function is then "well-arranged", i.e. the sequence of the factors is inviolable. I will not enter into the general rule for symmetricalising here;[10] the idea, if I understand it aright, is that H^{kl} is to be a *diagonal matrix*, and in other respects the normalised function, regarded as one of ordinary analysis, is to be identical with the one originally given.[11] We will satisfy these demands in a direct manner.

Then the authors postulate that the *matrices* q_l^{ik}, p_l^{ik} shall satisfy an infinite system of equations, as "equations of motion", and to begin with they write this system as follows:

$$
\left.
\begin{aligned}
\left(\frac{dq_l}{dt}\right)^{ik} &= \left(\frac{\partial H}{\partial p_l}\right)^{ik} \\
\left(\frac{dp_l}{dt}\right)^{ik} &= \left(-\frac{\partial H}{\partial q_l}\right)^{ik}
\end{aligned}
\right\}
\quad
\begin{aligned}
&l = 1, 2, 3, \ldots n \\
&i, k = 1, 2, 3, \ldots \text{ad.inf.}
\end{aligned}
\tag{18}
$$

The upper pair of indices signifies, as before in F^{ki} the respective element of the matrix belonging to the well-arranged function in question. The meaning of the partial differential coefficient on the right-hand side has just been explained, but *not* that of the $\frac{d}{dt}$ appearing on the left. By it the authors signify the following. *It* is to *give* a series of *numbers*

$$
\nu_1, \nu_2, \nu_3, \nu_4, \ldots \text{ ad. inf.,} \tag{19}
$$

such that the above equations are fulfilled, when to the $\frac{d}{dt}$ is ascribed the meaning: multiplication of the (ik)th matrix

---

[10] "Quantum Mechanics I." p. 873 *et seq.*

[11] The *stricter* postulation – "shall yield the same quantum-mechanical equations of motion" – I consider too narrow. It arises, in my opinion, from the fact that the authors confine themselves to *power products with regard also to the q_k's* – which is unnecessary.

element by $2\pi\sqrt{-1}(\nu_l - \nu_k)$. Thus, in particular,

$$
\begin{cases}
\left(\dfrac{dq_l}{dt}\right)^{ik} = 2\pi\sqrt{-1}(\nu_i - \nu_k)q_l^{ik}; \\[3mm]
\left(\dfrac{dp_l}{dt}\right)^{ik} = 2\pi\sqrt{-1}(\nu_i - \nu_k)p_l^{ik}.
\end{cases}
\tag{20}
$$

The series of numbers (19) is not defined in any way before-hand, but together with the matrix elements q_l^{ik}, p_l^{ik}, they form the numerical unknowns of the system of equations (18). The latter assumes the form

$$
\begin{cases}
(\nu_i - \nu_k)q_l^{ik} = \dfrac{1}{h}(Hq_l - q_l H) \\[3mm]
(\nu_i - \nu_k)p_l^{ik} = \dfrac{1}{h}(Hp_l - p_l H)
\end{cases}
\tag{18$'$}
$$

when we utilise the explanation of the symbols (20), and the calculating rules (14) and (15), and take account of (12).

We must thus satisfy *this* system of equations, and we have no means at our disposal, other than the suitable choice of the orthogonal system (4), which intervenes in the formation of the matrices. I now assert the following:

1. The equations (18$'$) will in general be satisfied if we choose as the orthogonal system the *proper functions* of the natural boundary value problem of the following partial differential equation,

$$
-[H, \psi] + E\psi = 0.
\tag{21}
$$

ψ is the unknown function of $q_1, q_2, \ldots, q_n$; E is the proper value parameter. Of course, as density function, $\rho(x)$ appears that function of $q_1, q_2, \ldots, q_n$, by which equation (21) must be multiplied in order to make it self-adjoint. The quantities ν_i are found to be equal to the proper values E_i divided by h. H^{kl} becomes a diagonal matrix, with $H^{kk} = E_k$.

2. If the symmetricalising of the function H has been effected *in a suitable way* – the process of symmetricalising, in my opinion, has not hitherto been defined uniquely – then (21) *is identical with the wave equation which is the basis of my wave mechanics.*[12]

Assertion 1 is almost directly evident, if we provisionally lay aside the questions whether equation (21) gives rise at all to an intelligible boundary value problem with the domain of entire q-space, and whether it can always be made self-adjoint through multiplication by a suitable function, etc. These questions are largely settled under heading 2. For now we have, according to (21) and the definitions of proper values and functions,

$$[H, u_i] = E_i u_i, \tag{22}$$

and thus from (6) we get

$$\begin{cases} H^{kl} = \int \rho(x)u_k(x)[H, u_l(x)]dx = E_i \int \rho(x)u_k(x)u_l(x)dx \\ = 0 \quad \text{for} \quad l \neq k \\ = E_l \quad \text{for} \quad l = k, \end{cases} \tag{23}$$

and, for example,

$$\begin{cases} (Hq_l)^{ik} = \sum_m H^{im}q_l^{mk} = E_i q_l^{ik} \\ (q_l H)^{ik} = \sum_m q_l^{im} H^{mk} = E_k q_l^{ik} \end{cases} \tag{24}$$

so that the right-hand side of the first equation of (18′) takes the value

$$\frac{E_i - E_k}{h} q_l^{ik}. \tag{25}$$

Similarly for the second equation. Thus everything asserted under 1 is proved.

---

[12]Equation (18″), Part II.

Let us turn now to assertion 2, which is, that there is
agreement between the negatively taken operator of the Hamilton function (suitably symmetricalised) and the wave operator of wave mechanics. I will first illustrate by a simple example why the process of symmetricalisation seems to me to
be, *in the first instance*, not unique. Let, for *one* degree of
freedom, the *ordinary* Hamilton function be

$$H = \frac{1}{2}\left(p^2 + q^2\right).\tag{26}$$

Then it is admitted that we can take this function, just as it
stands, unchanged, over to "quantum mechanics" as a "well-
arranged" function. But we can also, and seemingly indeed
with as much right *to begin with*, apply the well-arranged
function

$$H = \frac{1}{2}\left(\frac{1}{f(q)}pf(q)p + q^2\right),\tag{27}$$

where $f(q)$ is a function arbitrary within wide limits. $f(q)$
would appear in this case as a "density function" $\rho(x)$. (26)
is quite evidently just a special case of (27), and the question
arises, whether (and how) it is at all possible to distinguish
the special case we are concerned with, i.e. for more complicated H-functions. Confining ourselves to power products
only of the q_k's (where we could then simply prohibit the "pro-
duction of denominators") would be most inconvenient just in
the most important applications. Besides, I believe that does
not lead to correct symmetricalisation.

For the convenience of the reader, I will now give again
a short derivation of the wave equation in a form suited to
the present purpose, confining myself to the case of classical mechanics (without relativity and magnetic fields). Let,
therefore,

$$H = T(q_k, p_k) + V(q_k),\tag{28}$$

T being a quadratic form in the p_k's. Then the wave equation can be deduced[13] from the following variation problem,

$$
\begin{cases}
\delta J_1 = \delta \int \left\{ \dfrac{h^2}{4\pi^2} T\left(q_k, \dfrac{\partial\psi}{\partial q_k}\right) + \psi^2 V(q_k) \right\} \Delta_p^{-\frac{1}{2}}\, dx = 0, \\[2mm]
\text{with the subsidiary condition} \\[2mm]
J_2 = \int \psi^2 \Delta_p^{-\frac{1}{2}}\, dx = 1.
\end{cases}
\tag{29}
$$

As above, $\int dx$ stands for $\int \ldots \int dq_1 \ldots dq_n$; $\Delta_p^{-\frac{1}{2}}$ is *the reciprocal of the square root of the discriminant* of the quadratic form T. *This factor must not be omitted,* because otherwise the whole process would not be invariant for point transformations of the q's! By all means another explicit function of the q's might appear as a factor, i.e. a function which would be invariant for a point transformation of the q's. (For Δ_p, as is known, this is not the case. Otherwise we *could* omit $\Delta_p^{-\frac{1}{2}}$, if this extra function was given the value $\Delta_p^{\frac{1}{2}}$.)

If we indicate the derivative of T with respect to that argument, which originally was p_k, by the suffix p_k, we obtain, as the *result of the variation,*

$$
\begin{cases}
0 \ = \dfrac{1}{2}\left(\delta J_1 - E\delta J_2\right) \\[2mm]
\quad = \int \left\{ -\dfrac{h^2}{8\pi^2} \sum_k \dfrac{\partial}{\partial q_k}\left[\Delta_p^{-\frac{1}{2}} T_{p_k}\left(q_k, \dfrac{\partial\psi}{\partial q_k}\right)\right] \right. \\[2mm]
\quad\quad \left. + \left(V(q_k) - E\right)\Delta_p^{-\frac{1}{2}}\psi \right\}\delta\psi dx;
\end{cases}
\tag{30}
$$

the Eulerian variation equation thus runs:

$$
\frac{h^2}{8\pi^2}\Delta_p^{\frac{1}{2}} \sum_k \frac{\partial}{\partial q_k}\left\{ \Delta_p^{-\frac{1}{2}} T_{p_k}\left(q_k, \frac{\partial\psi}{\partial q_k}\right)\right\} - V(q_k)\psi + E\psi = 0.
\tag{31}
$$

---

[13]Equations (23) and (24) of Part I.

It is not difficult to see that this equation has the form of (21) if we remember our law connecting the operators, and consider

$$T(q_k, p_k) = \frac{1}{2} \sum_k p_k T_{p_k}(q_k, p_k) \qquad (32)$$

the Eulerian equation for homogeneous functions, applied to the quadratic form T. In actual fact, if we detach the operator from the left side of (31), with the proper value term $E\psi$ removed, and replace in it $\frac{h}{2\pi\sqrt{-1}}\frac{\partial}{\partial q_k}$ by p_k, then according to (32) we obtain the negatively taken Hamilton function (28). Thus the process of variation has given quite automatically a uniquely defined "symmetricalisation" of the operator, which makes it self-adjoint (except possibly for a common factor) and makes it invariant for point transformations, and which I would like to maintain, as long as there are no definite reasons for the appearance under the integrals (29) of the additional factor, already[14] mentioned as possible, and for a definite form of the latter.

Hence the solution of the whole system of matrix equations of Heisenberg, Born, and Jordan is reduced to the natural boundary value problem of a linear partial differential equation. If we have solved the boundary value problem, then by the use of (6) we can calculate by differentiations and quadratures every matrix element we are interested in.

As an illustration of what is to be understood by the *natural* boundary value problem, i.e. by the natural boundary conditions at the natural boundary of configuration space, we may refer to the worked examples.[15] It invariably turns out that the natural infinitely distant boundary forms a singularity of the differential equation and only allows of the one boundary condition – "remaining finite." This seems to be a general characteristic of those micro-mechanical problems

---

[14]Cf. also *Ann. d. Phys.* 79, p. 362 and p. 510 (i.e. Parts I. and II.).
[15]In Parts I. and II. of this collection.

with which the theory in the first place is meant to deal. If the domain of the position coordinates is artificially limited (example: a molecule in a "vessel"), then an essential allowance must be made for this limitation by the introduction of suitable potential energies in the well-known manner. Also the *vanishing* of the proper functions at the boundary generally occurs to an adequate degree, even if relations among *certain* of the integrals (6) are present, which necessitate a special investigation, and into which I will not enter at present. (It has to do with those matrix elements in the Kepler problem which, according to Heisenberg, correspond to the transition from one hyperbolic orbit to another.)

I have confined myself here to the case of classical mechanics, without magnetic fields, because the relativistic magnetic generalisation does not seem to me to be sufficiently clear yet. But we can scarcely doubt that the complete parallel between the two new quantum theories will still stand when this generalisation is obtained.

We conclude with a general observation on the whole formal apparatus of §§ 2, 3, and 4. The basic orthogonal system was regarded as an absolutely *discrete* system of functions. Now, in the most important applications this is *not* the *case*. Not only in the hydrogen atom but also in heavier atoms the wave equation (31) must possess a continuous proper value spectrum as well as a line spectrum. The former manifests itself, for example, in the continuous *optical* spectra which adjoin the limit of the series. It appeared better, provisionally, not to burden the formulae and the line of thought with this generalisation, though it is indeed indispensable. The chief aim of this paper is to work out, in the clearest manner possible, the formal connection between the two theories, and this is certainly not changed, in any essential point, by the appearance of a continuous spectrum. An important precaution that we have always observed is not to postulate, without further investigation, the convergence of the development in a series of proper functions. This precaution is especially demanded by the *accumulation of the proper values at a finite point* (viz. the limit of the series). This accumulation is most intimately connected with the appearance of the continuous spectrum.

§ 5. Comparison of the Two Theories. Prospect of a Classical Understanding of the Intensity and Polarisation of the Emitted Radiation

If the two theories – I might reasonably have used the singular – should[16] be tenable in the form just given, i.e. for more complicated systems as well, then every discussion of the superiority of the one over the other has only an illusory object, in a certain sense. For they are completely equivalent from the mathematical point of view, and it can only be a question of the subordinate point of convenience of calculation.

Today there are not a few physicists who, like Kirchhoff and Mach, regard the task of physical theory as being merely a mathematical description (as economical as possible) of the empirical connections between observable quantities, i.e. a description which reproduces the connection, as far as possible, without the intervention of unobservable elements. On this view, mathematical equivalence has almost the same meaning as physical equivalence. In the present case there might perhaps appear to be a certain superiority in the ma-

---

[16]There is a special reason for leaving this question open. The two theories initially take the energy function over from ordinary mechanics. Now in the cases treated the *potential* energy arises from the interaction of particles, of which perhaps *one*, at least, may be regarded in wave mechanics also as forming a point, on account of its great mass (cf. A. Einstein, *Berl. Ber.* 1925, p. 10). We must take into account the possibility that it is no longer permissible to take over from ordinary mechanics the statement for the potential energy, if both "point charges" are really extended states of vibration, which penetrate each other.

trix representation because, through its stifling of intuition, it does not tempt us to form space-time pictures of atomic processes, which must perhaps remain uncontrollable. In this connection, however, the following *supplement* to the proof of equivalence given above is interesting. The equivalence *actually* exists, and it also exists *conversely*. Not only can the matrices be constructed from the proper functions as shown above, but also, conversely, the functions can be constructed from the numerically given matrices. Thus the functions do not form, as it were, an *arbitrary* and *special* "fleshly clothing" for the bare matrix skeleton, provided to pander to the need for intuitiveness. This really would establish the superiority of the matrices, from the epistemological point of view. We suppose that in the equations

$$q_l^{ik} = \int u_i(x)u_k(x)dx \tag{33}$$

the *left-hand* sides are given numerically and the functions $u_i(x)$ are to be found. (N.B. – The "density function" is omitted for simplicity; the $u_i(x)$'s *themselves* are to be orthogonal functions for the present.) We may then calculate by matrix multiplication (without, by the way, any "revolving", i.e. integration by parts) the following integrals,

$$\int P(x)u_i(x)u_k(x)dx, \tag{34}$$

where $P(x)$ signifies *any* power product of the q_l's. The totality of these integrals, when i and k are fixed, forms what is called the totality of the *"moments"* of the function $u_i(x)u_k(x)$. And it is known that, under very general assumptions, a function is determined uniquely by the totality of its moments. So all the products $u_i(x)u_k(x)$ are uniquely fixed, and thus also the squares $u_i(x)^2$, and therefore also $u_i(x)$ itself. The only arbitrariness lies in the supplementary detachment of the density function $\rho(x)$, e.g. $r^2 \sin \theta$ in polar co-ordinates. No false step is to be feared there, certainly not so far as *epistemology* is concerned.

Moreover, the validity of the thesis that mathematical and physical equivalence mean the same thing, must itself be qualified. Let us think, for example, of the two expressions for the electrostatic energy of a system of charged conductors, the space integral $\frac{1}{2} \int E^2 d\tau$ and the sum $\frac{1}{2} \sum e_i V_i$ taken over the conductors. The two expressions are completely equivalent in electrostatics; the one may be derived from the other by integration by parts. Nevertheless we intentionally prefer the first and say that *it* correctly localises the energy in space. In the domain of electrostatics this preference has admittedly no justification. On the contrary, it is due simply to the fact that the first expression remains useful in electrodynamics also, while the second does not.

We cannot yet say with certainty to which of the two new quantum theories preference should be given, from *this* point of view. As the natural advocate of one of them, I will not be blamed if I frankly – and perhaps not wholly impartially – bring forward the arguments in its favour.

Leaving aside the special optical questions, the problems which the course of development of atomic dynamics brings up for consideration are presented to us by experimental physics in an eminently intuitive form; as, for example, how two colliding atoms or molecules rebound from one another, or how an electron or α-particle is diverted, when it is shot through an atom with a given velocity and with the initial path at a given perpendicular distance from the nucleus. In order to treat such problems more particularly, it is necessary to survey clearly the transition between macroscopic, perceptual mechanics and the micro-mechanics of the atom. I have lately[17] explained how I picture this transition. Micromechanics appears as a refinement of macro-mechanics, which

---

[17]Part II.

is necessitated by the geometrical and mechanical smallness of the objects, and the transition is of the same nature as that from geometrical to physical optics. The latter is demanded as soon as the wave length is no longer very great compared with the dimensions of the objects investigated or with the dimensions of the space inside which we wish to obtain more accurate information about the light distribution. To me it seems extraordinarily difficult to tackle problems of the above kind, as long as we feel obliged on epistemological grounds to repress intuition in atomic dynamics, and to operate only with such abstract ideas as transition probabilities, energy levels, etc.

An especially important question – perhaps the cardinal question of all atomic dynamics – is, as we know, that of the *coupling* between the dynamic process in the atom and the electromagnetic field, or whatever has to appear in the place of the latter. Not only is there connected with this the whole complex of questions of dispersion, of resonance- and secondary-radiation, and of the natural breadth of lines, but, in addition, the specification of certain quantities in atomic dynamics, such as emission frequencies, line intensities, etc., has only a mere dogmatic meaning until this coupling is described mathematically in some form or other. Here, now, the matrix representation of atomic dynamics has led to the conjecture that in fact the electromagnetic field also *must* be represented otherwise, namely, by matrices, so that the coupling may be mathematically formulated. Wave mechanics shows we are not compelled to do this in any case, for the mechanical field scalar (which I denote by ψ) is perfectly capable of entering into the unchanged Maxwell-Lorentz equations between the electromagnetic field vectors, as the "source" of the latter; just as, conversely, the electrodynamic potentials enter into the coefficients of the wave equation, which defines the field scalar.[18] In any case, it is worth while *attempting* the representation of the coupling in such a way that we bring into the unchanged Maxwell-Lorentz equations as *four-current* a four-dimensional vector, which has been suitably derived from the mechanical field scalar of the electronic motion (perhaps through the medium of the field vectors themselves, or the potentials). There even exists a hope that we can represent the wave equation for ψ equally well as a consequence of the Maxwell-Lorentz equations, namely, as an equation of continuity for electricity. The difficulty in regard to the problem of *several* electrons, which mainly lies in the fact that ψ is a function in *configuration* space, not in real space, must be mentioned. Nevertheless I would like to discuss the one-electron problem a little further, showing that it may be possible to give an extraordinarily clear interpretation of intensity and polarisation of radiation in this manner.

Let us consider the picture, on the wave theory, of the hydrogen atom, when it is in such a state that the field scalar ψ is given by a series of discrete proper functions, thus:

$$\psi = \sum_k c_k u_k(x) e^{\frac{2\pi\sqrt{-1}}{h} E_k t} \tag{35}$$

(x stands here for *three* variables, e.g. r, θ, ϕ; the c_k's are taken as real and it is correct to take the real part). We now

---

[18]Similar ideas are expressed by K. Lanczos in an interesting note that has just appeared (*Ztschr. f. Phys.* 35, p. 812, 1926). This note is also valuable as showing that Heisenberg's atomic dynamics is capable of a continuous interpretation as well. However, Lanczos' work has fewer points of contact with the present work than at first it was thought to have. The determination of his formal system, which was provisionally left quite indefinite, is *not* to be sought by following the idea that in some way the symmetrical nucleus $K(s, \sigma)$ of Lanczos can be identified with the *Green's function* of our wave equation (21) or (31). For this Green's function, if it exists, has the quantum levels themselves as proper values. On the other hand, it is required that Lanczos' function should have the *reciprocals* of the quantum levels as proper values.

make the *assumption* that the space density of electricity is given by the real part of

$$\psi \frac{\partial \bar{\psi}}{\partial t}. \tag{36}$$

The bar is to denote the conjugate complex function. We then calculate for the space density,

space density =

$$= 2\pi \sum_{(k,m)} c_k c_m \frac{E_k - E_m}{h} u_k(x) u_m(x) \sin \frac{2\pi t}{h} (E_m - E_k),$$

$$\tag{37}$$

where the sum is to be taken once only over every combination (k, m). Only term *differences* enter (37) as frequencies. The former are so low that the length of the corresponding ether wave is large compared with atomic dimensions, that is, compared with the region within which (37) is markedly different from zero.[19] The radiation can therefore be estimated simply by the *dipole moment* which according to (37) the whole atom possesses. We multiply (37) by a Cartesian coordinate q_l and by the "density function" $\rho(x)$, ($r^2 \sin \theta$ in the present case) and integrate over the whole space. According to (13), we get for the component of the dipole moment in the direction q_l

$$M q_l = 2\pi \sum_{(k,m)} c_k c_m q_l^{km} \frac{E_k - E_m}{h} \sin \frac{2\pi t}{h} (E_m - E_k), \quad (38)$$

Thus we really get a "Fourier development" of the atom's electric moment, in which only term *differences* appear as frequencies. The Heisenberg matrix elements q_l^{km} come into the coefficients in such a manner that their cooperating influence on the intensity and polarisation of the part of the radiation

---

[19] *Ann. d. Phys.* 79, p. 371, 1926, i.e. beginning of § 2, Part I. here.

concerned is completely intelligible on the grounds of classical electrodynamics.

The present sketch of the mechanism of radiation is far from completely satisfactory and is in no way final. Assumption (36) makes use, somewhat freely, of complex calculation, in order to put to one side undesired components of vibration whose radiation cannot be investigated at all in the simple way used for the dipole moment of the entire atom, because the corresponding ether wave lengths (about 0.01Å) lie far below atomic dimensions. Moreover, if we integrate over all space, then by (5) the space density (37) gives zero and not, as is required, a finite value, independent of the time, which requires to be normalised to the electronic charge. In conclusion, for completeness, account should be taken of magnetic radiation, since if there is a spatial distribution of electric currents, radiation is possible without the appearance of an electric moment, e.g. with a frame aerial.

Nevertheless it appears to be a well-founded hope that a real understanding of the nature of emitted radiation will be obtained on the basis of one of the two very similar analytical mechanisms which have been sketched here.

(Received March 18, 1926).

Quantisation as a Problem of Proper Values (Part III)

Perturbation Theory, with Application to the Stark Effect of the Balmer Lines

(*Annalen der Physik* (4), vol. 80, 1926)

Introduction. Abstract

As has already been mentioned at the end of the preceding paper,[1] the available range of application of the proper value theory can by comparatively elementary methods be considerably increased beyond the "directly soluble problems"; for proper values and functions can readily be approximately determined for *such* boundary value problems as are sufficiently closely related to a directly soluble problem. In analogy with ordinary mechanics, let us call the method in question the *perturbation* method. It is based upon the important *property of continuity* possessed by proper values and functions,[2] principally, for our purpose, upon their *continuous* dependence on the *coefficients* of the differential equation, and less upon the extent of the domain and on the boundary conditions, since in our case the domain ("entire q-space") and the boundary conditions ("remaining finite") are generally the same for the unperturbed and perturbed problems.

The method is essentially the same as that used by Lord

---

[1] Last two paragraphs of Part II.

[2] Courant-Hilbert, chap. vi. §§ 2, 4, p. 337.

Rayleigh in investigating[3] the vibrations of a string with small *inhomogeneities* in his *Theory of Sound* (2nd edit., vol. i., pp. 115-118, London, 1894). This was a particularly simple case, as the differential equation of the unperturbed problem had *constant* coefficients, and only the perturbing terms were arbitrary functions along the string. A complete generalisation is possible not merely with regard to these points, but also for the specially important case of *several* independent variables, i.e. for *partial* differential equations, in which *multiple proper values* appear in the unperturbed problem, and where the addition of a perturbing term causes the *splitting up* of such values and is of the greatest interest in well-known spectroscopic questions (Zeeman effect, Stark effect, Multiplicities). In the development of the perturbation theory in the following Section I., which really yields nothing new to the mathematician, I put less value on generalising to the *widest possible extent* than on bringing forward the very simple rudiments in the clearest possible manner. From the latter, any desired generalisation arises almost automatically when needed. In Section II., as an example, the Stark effect is discussed and, indeed, by *two* methods, of which the *first* is analogous to Epstein's method, by which he first solved[4] the problem on the basis of classical mechanics, supplemented by quantum conditions, while the *second*, which is much more general, is analogous to the method of secular perturbations.[5] The *first* method will be utilised to show that in wave mechanics also the perturbed problem can be "separated" in *parabolic* coordinates, and the perturbation theory will first be applied to the ordinary differential equations into which the original vibration equation is split up. The theory thus merely takes over the task which on the old theory devolved on Sommerfeld's elegant complex integration for the calculation of the quan-

---

[3]Courant-Hilbert, chap. v. § 5, 2, p. 241.

[4]P. S. Epstein, *Ann. d. Phys.* 50, p. 489, 1910.

[5]N. Bohr, *Kopenhagener Akademie* (8), IV., 1, 2, p. 69 *et seq.*, 1918.

tum integrals.[6] In the *second* method, it is found that in the case of the Stark effect an exact separation coordinate system exists, quite by accident, for the perturbed problem also, and the perturbation theory is applied directly to the *partial* differential equation. This latter proceeding proves to be more troublesome in wave mechanics, although it is theoretically superior, being more capable of generalisation.

Also the problem of the intensity of the components in the Stark effect will be shortly discussed in Section II. Tables will be calculated, which, as a whole, agree even better with experiment than the well-known ones calculated by Kramers with the help of the correspondence principle.[7]

The application (not yet completed) to the *Zeeman effect* will naturally be of much greater interest. It seems to be indissolubly linked with a correct formulation in the language of wave mechanics of the *relativistic* problem, because in the four-dimensional formulation the vector-potential automatically ranks equally with the scalar. It was already mentioned in Part I. that the relativistic hydrogen atom may indeed be treated without further discussion, but that it leads to "half-integral" azimuthal quanta, and thus contradicts experience. Therefore "something must still be missing". Since then I have learnt *what* is lacking from the most important publications of G. E. Uhlenbeck and S. Goudsmit,[8] and then from oral and written communications from Paris (P. Langevin) and Copenhagen (W. Pauli), viz., in the language of the theory of electronic orbits, the *angular momentum* of the electron round its axis, which gives it a *magnetic moment*. The utterances of these investigators, together with two highly significant pa-

---

[6]A. Sommerfeld, *Atombau*, 4th ed., p. 772.

[7]H. A. Kramers, *Kopenhagener Akademie* (8), III., 3, p. 287, 1919.

[8]G. E. Uhlenbeck and S. Goudsmit, *Physica*, 1925 ; *Die Naturwissenschaften*, 1926; *Nature*, 20th Feb., 1926; cf. also L. H. Thomas, *Nature*, 10th April, 1926.

pers by Slater[9] and by Sommerfeld and Unsold[10] dealing with the Balmer spectrum, leave no doubt that, by the introduction of the paradoxical yet happy conception of the spinning electron, the orbital theory will be able to master the disquieting difficulties which have latterly begun to accumulate (anomalous Zeeman effect; Paschen-Back effect of the Balmer lines; irregular and regular Róntgen doublets; analogy of the latter with the alkali doublets, etc.). We shall be obliged to attempt to take over the idea of Uhlenbeck and Goudsmit into wave mechanics. I believe that the latter is a very fertile soil for this idea, since in it the electron is not considered as a point charge, but as continuously flowing through space,[11] and so the unpleasing conception of a "rotating point-charge" is avoided. In the present paper, how- ever, the taking over of the idea is not yet attempted.

To the *third section*, as "mathematical appendix", have been relegated numerous uninteresting calculations – mainly quadratures of products of proper functions, required in the second section. *The formulae of the appendix are numbered (101), (102), etc.*

I. PERTURBATION THEORY

§ 1. A Single Independent Variable

Let us consider a linear, homogeneous, differential expression of the second order, which we may assume to be in self-adjoint form without loss of generality, viz.

$$L[y] = py'' + p'y' - qy. \tag{1}$$

y is the dependent function; p, p' and q are continuous functions of the independent variable x and $p \geq 0$. A dash denotes

---

[9] J. C. Slater, *Proc. Amer. Nat. Acad.* 11, p. 732, 1925.
[10] A. Sommerfeld and A. Unsold, *Ztschr. f. Phys.* 36, p. 259, 1926.
[11] Cf. last two pages of previous paper.

differentiation with respect to x (p' is therefore the derivative of p, which is the condition for self-adjointness).

Now let $\rho(x)$ be another continuous function of x, which never becomes negative, and also in general does not vanish. We consider the proper value problem of Sturm and Liouville,[12]

$$L[y] + E\rho y = 0. \tag{2}$$

It is a question, first, of finding all *those* values of the constant E ("proper values") for which the equation (2) possesses solutions $y(x)$, which are continuous and not identically vanishing within a certain domain, and which satisfy certain "boundary conditions" at the bounding points; and secondly of finding these solutions ("proper functions") themselves. In the cases treated in atomic mechanics, domain and boundary conditions are always "natural." The domain, for example, reaches from 0 to ∞, when x signifies the value of the radius vector or of an intrinsically positive parabolic coordinate, and the boundary conditions are in these cases: *remaining finite*. Or, when x signifies an azimuth, then the domain is the interval from 0 to 2π and the condition is: Repetition of the initial values of y and y' at the end of the interval ("periodicity").

It is only in the case of the periodic condition that *multiple*, viz. *double-valued*, proper values appear for *one* independent variable. By this we understand that to the same proper value belong *several* (in the particular case, two) linearly independent proper functions. We will now exclude this case for the sake of simplicity, as it attaches itself easily to the developments of the following paragraph. Moreover, to lighten the formulae, we will not expressly take into account in the notation the possibility that a "band spectrum" (i.e. a *continuum* of proper values) may be present when the domain extends to infinity.

---

[12]Cf. Courant-Hilbert, chap. v. § 5, 1, p. 238 *et seq.*

Let now $y = u_i(x)$, $i = 1, 2, 3, \ldots$, be the series of Sturm-Liouville proper functions; then the series of functions $u_i(x)\sqrt{\rho(x)}$, $i = 1, 2, 3, \ldots$, forms a *complete orthogonal system* for the domain; i.e. in the first place, if $u_i(x)$ and $u_k(x)$ are the proper functions belonging to the values E_i and E_k, then

$$\int \rho(x)u_i(x)u_k(x)dx = 0 \text{ for } i \neq k. \tag{3}$$

(Integrals without limits are to be taken over the domain, throughout this paper.) The expression "complete" signifies that an originally arbitrary continuous function is condemned to vanish identically, by the mere postulation that it must be orthogonal with respect to *all* the functions $u_i(x)\sqrt{\rho(x)}$. (More shortly: "There exists no further orthogonal function for the system.") We can and will always regard the proper functions $u_i(x)$ in all general discussions as "normalised", i.e. we imagine the constant factor, which is still arbitrary in each of them on account of the homogeneity of (2), to be defined in *such a way* that the integral (3) takes the value unity for $i = k$. Finally we again remind the reader that the proper values of (2) are certainly all real.

Let now the proper values E_i and functions $u_i(x)$ be *known*. Let us, from now on, direct our attention specially to a *definite* proper value, E_k say, and the corresponding function $u_k(x)$, and ask how these alter, when we do not alter the problem in any way other than by adding to the left-hand side of (2) a small "perturbing term", which we will initially write in the form

$$- \lambda r(x)y. \tag{4}$$

In this λ is a small quantity (the perturbation parameter), and $r(x)$ is an arbitrary continuous function of x. It is therefore simply a matter of a slight alteration of the coefficient q in the differential expression (1). From the continuity properties of the proper quantities, mentioned in the introduction, we now know that the altered Sturm-Liouville problem

$$L[y] - \lambda r y + E \rho y = 0 \qquad (2')$$

must have, in any case for a sufficiently small λ, proper quantities in the near neighbourhood of E_k and u_k, which we may write, by way of trial, as

$$E_k^* = E_k + \lambda \epsilon_k; \quad u_k^* = u_k(x) + \lambda v_k(x). \qquad (5)$$

On substituting in equation $(2')$, remembering that u_k satisfies (2), neglecting λ^2 and cutting away a factor λ we get

$$L[v_k] + E_k \rho v_k = (r - \epsilon_k \rho) u_k. \qquad (6)$$

For the defining of the perturbation v_k of the proper *function*, we thus obtain, as a comparison of (2) and (6) shows, a *non-homogeneous* equation, which belongs precisely to *that* homogeneous equation which is satisfied by our unperturbed proper function u_k (for in (6) the special proper value E_k stands in place of E). On the right-hand side of this non-homogeneous equation occurs, in addition to known quantities, the still unknown perturbation ϵ_k of the proper *value*.

This occurrence of ϵ_k serves for the calculation of this quantity *before* the calculation of v_k. It is known that the non-homogeneous equation – and this is *the starting point of the whole perturbation theory* — for a proper *value* of the homogeneous equation possesses a solution *when*, and *only when*, its right-hand side is *orthogonal*[13] to the allied proper function (to all the allied functions, in the case of multiple proper values). (The physical interpretation of this mathematical theorem, for the vibrations of a string, is that if the force is in resonance with a proper vibration it must be distributed in a very special way over the string, namely, so that it does no work in the vibration in question; otherwise the

---

[13]Cf. Courant-Hilbert, chap. v. § 10, 2, p. 277.

amplitude grows beyond all limits and a stationary condition is impossible.)

The right-hand side of (6) must therefore be orthogonal to u_k, i.e.

$$\int (r - \epsilon_k \rho) u_k^2 dx = 0, \tag{7}$$

or

$$\epsilon_k = \frac{\displaystyle\int r u_k^2 dx}{\displaystyle\int \rho u_k^2 dx}, \tag{7'}$$

or, if we imagine u_i already normalised, then, more simply,

$$\epsilon_k = \int r u_k^2 dx. \tag{7''}$$

This simple formula expresses the perturbation of the proper value (of first order) in terms of the perturbing function $r(x)$ and the unperturbed proper function $u_k(x)$. If we consider that the proper value of our problem signifies mechanical energy or is analogous to it, and that the proper function u_k is comparable to "motion with energy E_k", then we see in (7″) the complete parallel to the well-known theorem in the perturbation theory of classical mechanics, viz. the perturbation of the energy, to a first approximation, is equal to the perturbing function, averaged over the unperturbed motion. (It may be remarked in passing that it is as a rule sensible, or at least aesthetic, to throw into bold relief the factor $\rho(x)$ in the integrands of all integrals taken over the entire domain. If we do this, then, in integral (7″), we must speak of $\frac{r(x)}{\rho(x)}$ and not $r(x)$ as the perturbing function, and make a corresponding change in the expression (4). Since the point is quite unimportant, however, we will stick to the notation already chosen.)

We have yet to define $v_k(x)$, the perturbation of the proper *function*, from (6). We solve[14] the non-homogeneous equation by putting for v_k a series of proper functions, viz.

$$v_k(x) = \sum_{i=1}^{\infty} \gamma_{ki} u_i(x), \tag{8}$$

and by developing the right-hand side, divided by $\rho(x)$, likewise in a series of proper functions, thus

$$\left(\frac{r(x)}{\rho(x)} - \epsilon_k \right) u_k(x) = \sum_{i=1}^{\infty} c_{ki} u_i(x), \tag{9}$$

where

$$\begin{cases} c_{ki} &= \int (r - \epsilon_k \rho) u_k u_i dx \\ &= \int r u_k u_i dx \quad \text{for } i \neq k \\ &= 0 \quad \text{for } i = k. \end{cases} \tag{10}$$

The last equality follows from (7). If we substitute from (8) and (9) in (6) we get

$$\sum_{i=1}^{\infty} \gamma_{ki}(L[u_i] + E_k \rho u_i) = \sum_{i=1}^{\infty} c_{ki} \rho u_i. \tag{11}$$

Since now u_i satisfies equation (2) with $E = E_i$, it follows that

$$\sum_{i=1}^{\infty} \gamma_{ki} \rho(E_k - E_i) u_i = \sum_{i=1}^{\infty} c_{ki} \rho u_i. \tag{12}$$

By equating coefficients on left and right, all the γ_{ki}'s, except γ_{kk}, are defined. Thus

$$\gamma_{ki} = \frac{c_{ki}}{E_k - E_i} = \frac{\int r u_k u_i dx}{E_k - E_i} \quad \text{for } i \neq k, \tag{13}$$

---

[14]Cf. Courant-Hilbert, chap. v. § 5, 1, p. 240, and § 10, p. 279.

while γ_{kk}, as may be understood, remains completely undefined. This indefiniteness corresponds to the fact that the
postulation of normalisation is still available for us for the
perturbed proper function. If we make use of (8) in (5) and
claim for $u_k^*(x)$ the same normalisation as for $u_k(x)$ (quantities of the order of λ^2 being neglected), then it is evident that
$\gamma_{kk} = 0$. Using (13) we now obtain for the *perturbed proper
function*

$$u_k^*(x) = u_k(x) + \lambda \sum_{i=1}^{\infty}{}' \frac{u_i(x) \int r u_k u_i dx}{E_k - E_i}. \qquad (14)$$

(The dash on the sigma denotes that the term $i = k$ has not
to be taken.) And the allied perturbed proper value is, from
the above,

$$E_k^* = E_k + \lambda \int r u_k^2 dx. \qquad (15)$$

By substituting in (2′) we may convince ourselves that (14)
and (15) do really satisfy the proper value problem to the
proposed degree of approximation. This verification is necessary since the development, assumed in (5), in *integral* powers
of the perturbation parameter is no necessary consequence of
continuity.

The procedure, here explained in fair detail for the simplest case, is capable of generalisation in many ways. In the
first place, we can of course consider the perturbation in a
quite similar manner for the second, and then the third order
in λ, etc., in each case obtaining first the next approximation
to the proper value, and then the corresponding approximation for the proper function. In certain circumstances it may
be advisable – just as in the perturbation theory of mechanics
– to regard the perturbation function itself as a power series
in λ, whose terms come into play one by one in the separate
stages. These questions are discussed exhaustively by Herr E.

Fues in work which is now appearing in connection with the application to the theory of *band spectra*.

In the second place, in quite similar fashion, we can consider also a perturbation of the term in y' of the differential operator (1) just as we have considered above the term $-qy$. The case is important, for the Zeeman effect leads without doubt to a perturbation of this kind – though admittedly in an equation with several independent variables. Thus the equation loses its self-adjoint form by the perturbation – not an essential matter in the case of a single variable. In a partial differential equation, however, this loss may result in the perturbed proper values no longer being real, though the perturbing term is real; and naturally also conversely, an imaginary perturbing term may have a real, physically intelligible perturbation as its consequence.

We may also go further and consider a perturbation of the term in y''. Indeed it is quite possible, in general, to add an arbitrary "infinitely small" linear[15] and homogeneous differential operator, even of higher order than the second, as the perturbing term and to calculate the perturbations in the same manner as above. In these cases, however, we would use with advantage the fact that the second and higher derivatives of the proper functions may be expressed by means of the differential equation itself, in terms of the zero and first derivatives, so that this general case may be reduced, in a certain sense, to the two special cases, first considered – perturbation of the terms in y and y'.

Finally, it is obvious that the extension to equations of order higher than the second is possible.

Undoubtedly, however, the most important generalisation is that to several independent variables, i.e. to partial differential equations. For *this* really is the problem in the general case, and only in exceptional cases will it be possible to split

---

[15]Even the limitation "linear" is not absolutely necessary.

up the disturbed partial differential equation, by the introduction of suitable variables, into separate differential equations, each only with one variable.

§ 2. Several Independent Variables (Partial Differential Equation)

We will represent the several independent variables in the formulae symbolically by the *one* sign x, and briefly write $\int dx$ (instead of $\int \ldots \int dx_1 dx_2 \ldots$) for an integral extending over the multiply-dimensioned *domain*. A notation of this type is already in use in the theory of integral equations, and has the advantage, here as there, that the structure of the formulae is not altered by the increased number of variables as such, but only by *essentially* new occurrences, which *may* be related to it.

Let therefore $L[y]$ now signify a self-adjoint *partial* linear differential expression of the second order, whose explicit form we do not require to specify; and further let $\rho(x)$ again be a positive function of the independent variables, which does not vanish in general. The postulation "self-adjoint" is *now* no longer unimportant, as the property cannot now be generally gained by multiplication by a suitably chosen $f(x)$, as was the case with *one* variable. In the particular differential expression of wave mechanics, however, this is still the case, as it arises from a variation principle.

According to these definitions or conventions, we can regard equation (2) of § 1,

$$L[y] + E\rho y = 0, \tag{2}$$

as the formulation of the Sturm-Liouville proper value problem in the case of several variables also. Everything said there about the proper values and functions, their orthogonality, normalisation, etc., as also *the whole perturbation theory there developed* – in short, the whole of § 1 – *remains valid without change, when* all the proper values are *simple,* if we use the abbreviated symbolism just agreed upon above. And only *one* thing does *not* remain valid, namely, that they *must* be simple.

Nevertheless, from the pure mathematical standpoint, the case when the roots are all distinct is to be regarded as the *general* case for several variables also, and multiplicity regarded as a special occurrence, which, it is admitted, *is the rule in applications,* on account of the specially simple and symmetrical structure of the differential expressions $L[y]$ (and the "boundary conditions") which appear. Multiplicity of the proper values corresponds to *degeneracy* in the theory of conditioned periodic systems and is therefore especially interesting for quantum theory.

A proper value E_k is called α-fold, when equation (2), for $E = E_k$, possesses not *one* but exactly α linearly independent solutions which satisfy the boundary conditions. We will denote these by

$$u_{k_1}, u_{k_2}, \ldots u_{k_\alpha}. \tag{16}$$

Then it is true that each of these α proper functions is *orthogonal* to each of the *other* proper functions belonging to *another* proper value (the factor $\rho(x)$ being included; cf. (3)). On the contrary, these α functions are *not* in general orthogonal *to one another,* if we merely postulate that they are α linearly independent proper functions for the proper value E_k, and nothing more. For then we can equally well replace them by α arbitrary, linearly independent, linear aggregates (with constant coefficients) of themselves. We may express this otherwise, thus. The series of functions (16) is initially *indefinite* to the extent of a linear transformation (with constant coefficients), involving a non-vanishing determinant, and such a transformation *destroys,* in general, the mutual orthogonality.

But through such a transformation this mutual orthogonality can always be *brought about,* and indeed in an infinite number of ways; the latter property arising because *orthogonal* transformation does *not* destroy the mutual orthogonality. We are now accustomed to include this simply in *normalisation*, that orthogonality is secured for *all* proper functions, even for those which belong to the *same* proper value. We will assume that our u_{ki}'s are already *normalised* in this way, and of course for *each* proper value. Then we must have

$$\begin{cases} \int\!\!\int \rho(x)u_{ki}(x)u_{k'i'}(x)dx = 0 \text{ when } (k,i) \neq (k',i') \\ \qquad\qquad = 1 \text{ when } k' = k, \text{ as well as } i' = i. \end{cases}$$
$$(17)$$

Each of the finite series of proper functions u_{ki}, obtained for *constant* k and *varying* i, is then only still indefinite to this extent, that it is subject to an *orthogonal* transformation.

We will now discuss, first in words, without using formulae, the consequences which follow when a perturbing term is added to the differential equation (2). The addition of the perturbing term will, in general, remove the above-mentioned symmetry of the differential equation, to which the multiplicity of the proper values (or of certain of them) is due. Since, however, the proper values and functions are *continuously* dependent on the coefficients of the differential equation, a small perturbation causes a group of α proper values, which lie close to one another and to E_k, to enter in place of the α-fold proper value E_k. The latter is *split up*. Of course, if the symmetry is not wholly destroyed by the perturbation, it may happen that the splitting up is not complete and that several proper values (still partly multiple) of, *in summa*, equal multiplicity merely appear in the place of E_k ("*partial* removal of degeneracy").

As for the perturbed proper *functions*, those α members which belong to the α values arising from E_k must evidently also on account of continuity lie infinitely near the unperturbed functions belonging to E_k, viz. u_{ki}; $i = 1, 2, 3, \ldots \alpha$. Yet we must remember that the last-named series of functions, as we have established above, is indefinite to the extent of an *arbitrary orthogonal transformation*. *One* of the infinitely numerous definitions, which may be applied to the series of functions, u_{ki}; $i = 1, 2, 3, \ldots \alpha$, will lie infinitely near the series of perturbed functions; and if the value E_k is completely split up, it will be a *quite definite one*! For to the separate simple proper values, into which the value is split up, there belong proper functions which are quite uniquely defined.

This unique particular specification of the *unperturbed* proper functions (which may fittingly be designated as the "approximations of zero order" for the *perturbed* functions), which is defined by the nature of the perturbation, will naturally *not* generally coincide with that definition of the unperturbed functions which we chanced to adopt to begin with. Each group of the latter, belonging to a definite α-fold proper value E_k, will have first to be submitted to an orthogonal substitution, defined by the kind of perturbation, before it can serve as the starting-point, the "zero approximation", for a more exact definition of the perturbed proper functions. *The defining of these orthogonal substitutions* – one for each multiple proper value – *is the only essentially new point* that arises because of the increased number of variables, or from the appearance of multiple proper values. The defining of these substitutions forms the exact counterpart to the finding of an approximate separation system for the perturbed motion in the theory of conditioned periodic systems. As we will see immediately, the definition of the substitutions can always be given in a theoretically simple way. It requires, for each α-fold proper value, merely the principal axes transformation of a quadratic form of α (and thus of a finite number of) variables.

When the substitution has once been accomplished, the calculation of the approximations of the *first* order runs almost word for word as in § 1. The sole difference is that the dash on the sigma in equation (14) must mean that in the summation *all* the proper functions belonging to the value E_k, i.e. *all* the terms whose denominators would vanish, must be left out. It may be remarked in passing that it is not at all necessary, in the calculation of *first* approximations, to have completed the orthogonal substitutions referred to for *all* multiple proper values, but it is sufficient to have done so for the value E_k, in whose splitting up we are interested. For the approximations of higher order, we admittedly require them all. In all other respects, however, these higher approximations are from the beginning carried out exactly as for simple proper values.

Of course it may happen, as was mentioned above, that the value E_k either generally or at the initial stages of the approximation, is not completely split up, and that multiplicities ("degeneracies") still remain. This is expressed by the fact that to the substitutions already frequently mentioned there still clings a certain indefiniteness, which either always remains, or is removed step by step in the later approximations.

Let us now represent these ideas by formulae, and consider as before the perturbation caused by (4), § 1,

$$- \lambda r(x)y, \tag{4}$$

i.e. we imagine the proper value problem belonging to (2) *solved*, and now consider the exactly corresponding problem (2′),

$$L[y] - \lambda ry + E\rho y = 0. \tag{2′}$$

We again fix our attention on a definite proper value E_k. Let (16) be a system of proper functions belonging to it, which we assume to be normalised and orthogonal to one another in the sense described above, but *not yet* fitted to the particular perturbation in the sense explained, because to find the substitution that leads *to this fitting* is precisely our chief task! In place of (5), § 1, we must now put for the perturbed quantities the following,

$$E_{kl}^* = E_k + \lambda\epsilon_l; \quad u_{kl}^*(x) = \sum_{i=1}^{\alpha} \kappa_{li}u_{ki}(x) + \lambda v_l(x) \quad (18)$$

$$(l = 1, 2, 3, \ldots \alpha),$$

wherein the v_l's are functions, and the ϵ_l's and the κ_{li}'s are systems of constants, which are still to be defined, but which we initially do not limit in any way, although we know that the system of coefficients κ_{li} must[16] form an orthogonal substitution. The index k should still be attached to the three types of quantity named, in order to indicate that the whole discussion refers to the kth proper value of the unperturbed problem. We refrain from carrying this out, in order to avoid the confusing accumulation of indices. The index k is to be assumed *fixed* in the whole of the following discussion, until the contrary is stated.

Let us select *one* of the perturbed proper functions and values by giving a definite value to the index l in (18), and let us substitute from (18) in the differential equation (2′) and arrange in powers of λ. Then the terms independent of λ disappear exactly as in § 1, because the unperturbed proper quantities satisfy equation (2), by hypothesis. Only terms containing the *first* power of λ remain, as we can strike out the others. Omitting a factor λ, we get

$$L[v_l] + E_k\rho v_l = \sum_{i+1}^{\alpha} \kappa_{li}(r - \epsilon_l\rho)u_{ki}, \quad (19)$$

and thus obtain again for the definition of the perturbation v_l of the *functions* a *non-homogeneous* equation, to which

---

[16]It follows from the general theory that the perturbed system of functions $u_{kl}^*(x)$ *must* be orthogonal if the perturbation completely removes the degeneracy, and *may* be assumed orthogonal although that is not the case.

corresponds as homogeneous equation the equation (2), with the particular value $E = E_k$, i.e. the equation satisfied by the set of functions u_{ki}; $i = 1, 2, 3, \ldots \alpha$. The form of the left side of equation (19) is independent of the index l.

On the right side occur ϵ_l and κ_{li}, the constants to be defined, and we are thus enabled to evaluate them, even *before* calculating v_l. For, in order that (19) should have a solution at all, it is necessary and sufficient that its right-hand side should be orthogonal to *all* the proper functions of the homogeneous equation (2) belonging to E_k. Therefore, we must have

$$\left\{ \begin{aligned} &\sum_{i=1}^{\alpha} \kappa_{li} \int (r - \epsilon_l \rho) u_{ki} u_{km} dx = 0 \\ &(m = 1, 2, 3, \ldots \alpha), \end{aligned} \right. \tag{20}$$

i.e. on account of the normalisation (17),

$$\left\{ \begin{aligned} &\kappa_{lm} \epsilon_l = \sum_{i=1}^{\alpha} \kappa_{li} \int r u_{ki} u_{km} dx \\ &(m = 1, 2, 3, \ldots \alpha), \end{aligned} \right. \tag{21}$$

If we write, briefly, for the *symmetrical* matrix of constants, which can be evaluated by quadrature,

$$\left\{ \begin{aligned} &\int r u_{ki} u_{km} dx = \epsilon_{im} \\ &(i, m = 1, 2, 3, \ldots \alpha), \end{aligned} \right. \tag{22}$$

then we recognise in

$$\left\{ \begin{aligned} &\kappa_{lm} \epsilon_l = \sum_{i=1}^{\alpha} \kappa_{li} \epsilon_{mi} \\ &(m = 1, 2, 3, \ldots \alpha) \end{aligned} \right. \tag{21'}$$

a system of α linear homogeneous equations for the calculation of the α constants κ_{lm}; $m = 1, 2, 3, \ldots \alpha$, where the perturbation ϵ_l of the proper value still occurs in the coefficients, and is itself unknown. However, this serves for the calculation of ϵ_l before that of the κ_{lm}'s. For it is known that the linear homogeneous system (21') of equations has solutions if, and only if, its determinant vanishes. This yields the following algebraic equation of degree a for ϵ_l:

$$
\begin{vmatrix}
\epsilon_{11} - \epsilon_l, & \epsilon_{12} & , & \cdots & \epsilon_{1\alpha} \\
\epsilon_{21} & , & \epsilon_{22} - \epsilon_l, & \cdots & \epsilon_{2\alpha} \\
\cdots & & \cdots & \cdots & \cdots \\
\cdots & & \cdots & \cdots & \cdots \\
\epsilon_{\alpha 1} & , & \epsilon_{\alpha 2} & , \cdots & \epsilon_{\alpha\alpha} - \epsilon_l
\end{vmatrix} = 0 \qquad (23)
$$

We see that the problem is completely identical with the transformation of the quadratic form in α variables, with coefficients ϵ_{mi}, to its principal axes. The "secular equation" (23) yields α roots for ϵ_l, the "reciprocal of the squares of the principal axes," which in general are different, and on account of the symmetry of the ϵ_{mi}'s *always real*. We thus get all the α perturbations of the proper values $(l = 1, 2, 3, \ldots \alpha)$ at the same time, and would have *inferred* the splitting up of an α-fold proper value into exactly α simple values, generally different, even had we not assumed it already, as fairly obvious. For *each* of these ϵ_l-values, equations (21') give a system of quantities $\kappa_{li}; i = 1, 2, 3, \ldots \alpha$, and, as is known, *only one* (apart from a general constant factor), provided all the ϵ_l's are really different. Further, it is known that the whole system of α^2 quantities κ_{li} forms an *orthogonal* system of coefficients, defining as usual, in the principal axes problem, the *directions* of the new coordinate axes with reference to the old ones. We may, and will, employ the undefined factors just mentioned to normalise the κ_{li}'s completely as "direction cosines", and this, as is easily seen, makes the perturbed proper functions u_{ki}^* turn out *normalised* again, according to (18), at least in the "zero approximation" (i.e. apart from the λ-terms).

If the equation (23) has multiple roots, then we have the

case previously mentioned, when the perturbation does not completely remove the degeneration. The perturbed equation has then multiple proper values also and the definition of the constants κ_{li} becomes partially arbitrary. This has no consequence other than that (as is *always* the case with multiple proper values) we *must* and *may* acquiesce, even *after* the perturbation is applied, in a system of proper functions which in many respects is still arbitrary.

The main task is accomplished with this transformation to principal axes, and we will often find it sufficient in the applications in quantum theory to define the proper values to a first and the functions to zero approximation. The evaluation of the constants κ_{li} and ϵ_{li} cannot be carried out always, since it depends on the solution of an algebraic equation of degree α. At the worst there are methods[17] which give the evaluation to any desired approximation by a rational process. We may thus regard these constants as known, and will now give the calculation of the functions to the *first* approximation, for the sake of completeness. The procedure is exactly as in § 1.

We have to solve equation (19) and to that end we write v_l as a series of the *whole set* of proper functions of (2),

$$v_l(x) = \sum_{(k'\,i')} \gamma_{l,k'i'} u_{k'i'}(x). \tag{24}$$

The summation is to extend with respect to k' from 0 to ∞, and, for each fixed value of k', for i' varying over the finite number of proper functions which belong to $E_{k'}$. (Now, for the first time, we take account of proper functions which do *not* belong to the α-fold value E_k we are fixing our attention on.) Secondly, we develop the right-hand side of (19), divided by $\rho(x)$, in a series of the entire set of proper functions,

$$\sum_{i=1}^{\alpha} \kappa_{li} \left(\frac{r}{\rho} - \epsilon_l \right) u_{ki} = \sum_{(k'\,i')} c_{l,k'i'} u_{k'i'}, \tag{25}$$

---

[17]Courant-Hilbert, chap. i. § 3. 3, p. 14.

wherein

$$
\begin{cases}
c_{l,k'i'} & = \displaystyle\sum_{i=1}^{\alpha} \kappa_{li} \int (r - \epsilon_l \rho) u_{ki} u_{k'i'} dx \\[2mm]
& = \displaystyle\sum_{i=1}^{\alpha} \kappa_{li} \int r u_{ki} u_{k'i'} dx \quad \text{for } k' \neq k \\[2mm]
& = 0 \quad\quad\quad\quad\quad\quad\quad \text{for } k' = k
\end{cases}
\tag{26}
$$

(the last two equalities follow from (17) and (20) respectively). On substituting from (24) and (25) in (19), we get

$$
\sum_{(k'\,i')} \gamma_{l,k'i'} \left(L[u_{k'i'}] + E_k \rho u_{k'i'} \right) = \sum_{(k'\,i')} c_{l,k'i'} \rho u_{k'i'}. \tag{27}
$$

Since $u_{k'i'}$ satisfies equation (2) with $E = E_k$, this gives

$$
\sum_{(k'\,i')} \gamma_{l,k'i'} \rho \left(E_k - E_{k'} \right) u_{k'i'} = \sum_{(k'\,i')} c_{l,k'i'} \rho u_{k'i'}. \tag{28}
$$

By equating coefficients on right and left, all the $\gamma_{l,k'i'}$'s are defined, with the exception of those in which $k' = k$. Thus

$$
\begin{aligned}
\gamma_{l,k'i'} &= \frac{c_{l,k'i'}}{E_k - E_{k'}} \\[2mm]
&= \frac{1}{E_k - E_{k'}} \sum_{i=1}^{\alpha} \kappa_{li} \int r u_{ki} u_{k'i'} dx \quad (\text{for } k' \neq k),
\end{aligned}
\tag{29}
$$

while those γ's for which $k' = k$ are of course not fixed by equation (19). This again corresponds to the fact that we have provisionally normalised the perturbed functions u_{kl}^*, of (18), only in the zero approximation (through the normalisation of the κ_{li}'s), and it is easily recognised again that we have to put the whole of the γ-quantities in question equal to zero, in order to bring about the normalisation of the u_{kl}^*'s even in the first approximation. By substituting from (29) in (24), and then from (24) in (18), we finally obtain for the *perturbed proper functions to a first approximation*

$$u_{kl}^*(x) = \sum_{i=1}^{\alpha} \kappa_{li} \left(u_{ki}(x) + \lambda {\sum_{(k'\,i')}}' \frac{u_{k'i'}(x)}{E_k - E_{k'}} \int r u_{ki} u_{k'i'} dx \right)$$

(30)

$$(l = 1, 2, \ldots, \alpha).$$

The dash on the second sigma indicates that *all* the terms with $k' = k$ are to be omitted. In the application of the formula for an arbitrary k, it is to be observed that the κ_{li}'s, as obviously also the multiplicity α of the proper value E_k, to which we have specially directed our attention, still depend on the index k, though this is not expressed in the symbols. Let us repeat here that the κ_{li}'s are to be calculated as a system of solutions of equations (21'), normalised so that the sum of the squares is unity, where the coefficients of the equations are given by (22), while for the quantity ϵ_l in (21'), *one* of the roots of (23) is to be taken. *This* root then gives the allied perturbed proper *value*, from

$$E_{kl}^* = E_k + \lambda \epsilon_l. \tag{31}$$

Formulae (30) and (31) are the generalisations of (14) and (15) of § 1.

It need scarcely be said that the extensions and generalisations mentioned at the end of § 1 can of course take effect here also. It is hardly worth the trouble to carry out these developments generally. We succeed best in any special case if we do not use ready-made formulae, but go directly by the simple fundamental principles, which have been explained, perhaps too minutely, in the present paper. I would only like to consider briefly the possibility, already mentioned at the end of § 1, that the equation (2) perhaps may lose (and indeed in the case of several variables irreparably lose), its self-adjoint character if the perturbing terms also contain derivatives of the unknown function. From general theorems we know that then the proper values of the perturbed equation no longer need to be real. We can illustrate this further. We can easily see, by carrying out the developments of this paragraph, that the elements of determinant (23) are *no longer symmetrical*, when the perturbing term contains derivatives. It is known that in this case the roots of equation (23) no longer require to be real.

The necessity for the expansion of certain functions in a series of proper functions, in order to arrive at the first or zero approximation of the proper values or functions, can become very inconvenient, and can at least complicate the calculation considerably in cases where an extended spectrum co-exists with the point spectrum and where the point spectrum has a limiting point (point of accumulation) at a finite distance. This is just the case in the problems appearing in the quantum theory. Fortunately it is often – perhaps always – possible, for the purpose of the perturbation theory, to free oneself from the generally very troublesome extended spectrum, and to develop the perturbation theory from an equation which does *not* possess such a spectrum, and whose proper values do *not* accumulate near a finite value, but grow beyond all limits with increasing index. We will become acquainted with an example in the next paragraph. Of course, this simplification is only possible when we are not interested in a proper value of the extended spectrum.

II. APPLICATION TO THE STARK EFFECT

§ 3. Calculation of Frequencies by the Method which corresponds to that of Epstein

If we add a potential energy $+eFz$ to the wave equation (5), Part I., of the Kepler problem, corresponding to the influence of an electric field of strength F in the positive z-

direction, on a negative electron of charge e, then we obtain the following wave equation for the Stark effect of the hydrogen atom,

$$\nabla^2 \psi + \frac{8\pi^2 m}{h^2} \left(E + \frac{e^2}{r} - eFz \right) \psi = 0, \qquad (32)$$

which forms the basis of the remainder of this paper. In § 5 we will apply the general perturbation theory of § 2 directly to this partial differential equation. Now, however, we will lighten our task by introducing space parabolic coordinates $\lambda_1, \lambda_2, \phi$, by the following equations,

$$\begin{cases} x = \sqrt{\lambda_1 \lambda_2} \cos \phi \\ \qquad + \\ y = \sqrt{\lambda_1 \lambda_2} \sin \phi \\ \qquad + \\ z = \frac{1}{2}(\lambda_1 - \lambda_2). \end{cases} \qquad (33)$$

λ_1 and λ_2 run from 0 to infinity; the corresponding coordinate surfaces are the two sets of confocal paraboloids of revolution, which have the origin as focus and the positive (λ_2) or negative (λ_1) z-axis respectively as axes. ϕ runs from 0 to 2π, and the coordinate surfaces belonging to it are the set of half planes limited by the z-axis. The relation of the coordinates is *unique*. For the functional determinant we get

$$\frac{\partial(x, y, z)}{\partial(\lambda_1, \lambda_2, \phi)} = \frac{1}{4}(\lambda_1 + \lambda_2). \qquad (34)$$

The *space element* is thus

$$dx\,dy\,dz = \frac{1}{4}(\lambda_1 + \lambda_2)d\lambda_1 d\lambda_2 d\phi. \qquad (35)$$

We notice, as consequences of (33),

$$x^2 + y^2 = \lambda_1 \lambda_2; \quad r^2 = x^2 + y^2 + z^2 = \left\{ \frac{1}{2}(\lambda_1 + \lambda_2) \right\}^2. \qquad (36)$$

The expression of (32) in the chosen coordinates gives, if we multiply by (34)[18] (to restore the self-adjoint form),

$$
\begin{cases}
\dfrac{\partial}{\partial\lambda_1}\left(\lambda_1\dfrac{\partial\psi}{\partial\lambda_1}\right) + \dfrac{\partial}{\partial\lambda_2}\left(\lambda_2\dfrac{\partial\psi}{\partial\lambda_2}\right) + \dfrac{1}{4}\left(\dfrac{1}{\lambda_1}+\dfrac{1}{\lambda_2}\right)\dfrac{\partial^2\psi}{\partial\phi^2} \\[2mm]
+\dfrac{2\pi^2 m}{h^2}\left[E(\lambda_1+\lambda_2)+2e^2-\tfrac{1}{2}eF(\lambda_1^2-\lambda_2^2)\right]\psi = 0.
\end{cases}
$$

$$(32')$$

Here we can again take – and this is the why and wherefore of all "methods" of solving linear partial differential equations – the function ψ as the product of three functions, thus,

$$\psi = \Lambda_1\Lambda_2\Phi, \tag{37}$$

each of which depends on only *one* coordinate. For these functions we get the ordinary differential equations

$$
\begin{cases}
\dfrac{\partial^2\Phi}{\partial\phi^2} = -n^2\Phi \\[2mm]
\dfrac{\partial}{\partial\lambda_1}\left(\lambda_1\dfrac{\partial\Lambda_1}{\partial\lambda_1}\right)+ \\[2mm]
+\dfrac{2\pi^2 m}{h^2}\left(-\tfrac{1}{2}eF\lambda_1^2+E\lambda_1+e^2-\beta-\dfrac{n^2h^2}{8\pi^2 m}\dfrac{1}{\lambda_1}\right)\Lambda_1 = 0, \\[2mm]
\dfrac{\partial}{\partial\lambda_2}\left(\lambda_2\dfrac{\partial\Lambda_2}{\partial\lambda_2}\right)+ \\[2mm]
+\dfrac{2\pi^2 m}{h^2}\left(-\tfrac{1}{2}eF\lambda_2^2+E\lambda_2+e^2+\beta-\dfrac{n^2h^2}{8\pi^2 m}\dfrac{1}{\lambda_2}\right)\Lambda_2 = 0,
\end{cases}
$$

$$(38)$$

wherein n and β are two further "proper value-like" constants of integration (in addition to E), still to be defined. By the

---

[18]So far as the actual details of the analysis are concerned, the simplest way to get (32'), or, in general, to get the wave equation for any special coordinates, is to transform not the wave equation itself, but the corresponding variation problem (cf. Part I. p. 12), and thus to obtain the wave equation afresh as an Eulerian variation problem. We are thus spared the troublesome evaluation of the *second* derivatives. Cf. Courant-Hilbert, chap. iv. § 7, p. 193.

choice of symbol for the first of these, we have taken into account the fact that the first of equations (38) makes it take integral values, if Φ and $\frac{\partial \Phi}{\partial \phi}$ are to be continuous and single-valued functions of the azimuth ϕ. We then have

$$\Phi = \begin{matrix} \sin \\ \cos \end{matrix} n\phi \qquad (39)$$

and it is evidently sufficient if we do not consider negative values of n. Thus

$$n = 0, 1, 2, 3, \ldots \qquad (40)$$

In the symbol used for the second constant β, we follow Sommerfeld (*Atombau*, 4th edit., p. 821) in order to make comparison easier. (Similarly, below, with A, B, C, D.) We treat the last two equations of (38) together, in the form

$$\frac{\partial}{\partial \xi}\left(\xi \frac{\partial \Lambda}{\partial \xi} \right) + \left(D\xi^2 + A\xi + 2B + \frac{C}{\xi} \right)\Lambda = 0, \qquad (41)$$

where

$$\left. \begin{matrix} D_1 \\ D_2 \end{matrix} \right\} = \mp \frac{\pi^2 meF}{h^2}, \quad A = \frac{2\pi^2 mE}{h^2},$$

$$\left. \begin{matrix} B_1 \\ B_2 \end{matrix} \right\} = \frac{\pi^2 m}{h^2}(e^2 \mp \beta), \quad C = -\frac{n^2}{4}, \qquad (42)$$

and the upper sign is valid for $\Lambda = \Lambda_1$, $\xi = \lambda_1$ and the lower one for $\Lambda = \Lambda_2$, $\xi = \lambda_2$. (Unfortunately, we have to write ξ instead of the more appropriate λ, to avoid confusion with the perturbation parameter λ of the general theory, §§ 1 and 2.)

   If we omit initially in (41) the Stark effect term $D\xi^2$, which we conceive as a perturbing term (limiting case for vanishing field), then this equation has the same general structure as equation (7) of Part I., and the domain is also the same, from 0 to ∞. The discussion is almost the same, word for word, and

shows that non-vanishing solutions, which, with their derivatives, are continuous and remain finite within the domain,
only exist if *either* $A > 0$ (extended spectrum, corresponding
to hyperbolic orbits) *or*

$$\frac{B}{\sqrt[+]{-A}} - \sqrt[+]{-C} = k + \frac{1}{2}; \quad k = 0, 1, 2, \ldots \tag{43}$$

If we apply this to the last two equations of (38) and distinguish the two-values by suffixes 1 and 2, we obtain

$$\begin{cases} \sqrt[+]{-A}(k_1 + \tfrac{1}{2} + \sqrt[+]{-C}) = B_1 \\ \sqrt[+]{-A}(k_2 + \tfrac{1}{2} + \sqrt[+]{-C}) = B_2. \end{cases} \tag{44}$$

By addition, squaring and use of (42) we find

$$A = -\frac{4\pi^4 m^2 e^4}{h^4 l^2} \quad \text{and} \quad E = -\frac{2\pi^2 m.e^4}{h^2 l^2} \tag{45}$$

These are the well-known Balmer-Bohr elliptic levels, where
as *principal quantum number* enters

$$l = k_1 + k_2 + n + 1. \tag{46}$$

We get the *discrete* term spectrum and the allied proper functions in a way *simpler* than that indicated, if we apply results already known in mathematical literature as follows. We
transform first the dependent variable Λ in (41) by putting

$$\Lambda = \xi^{\frac{n}{2}} u \tag{47}$$

and then the independent ξ by putting

$$2\xi\sqrt{-A} = \eta. \tag{48}$$

We find for u as a function of η the equation

$$\frac{d^2 u}{d\eta^2} + \frac{n+1}{\eta}\frac{du}{d\eta} + \left(\frac{D}{(2\sqrt[+]{-A})^3}\eta - \frac{1}{4} + \frac{B}{\sqrt{-A}}\frac{1}{\eta} \right) u = 0. \tag{41'}$$

128

This equation is very intimately connected with the polynomials named after Laguerre. In the mathematical appendix, it will be shown that the product of $e^{-\frac{x}{2}}$ and the nth derivative of the $(n+k)$th Laguerre polynomial satisfies the differential equation

$$y'' + \frac{n+1}{x}y' + \left(-\frac{1}{4} + \left(k + \frac{n+1}{2}\right)\frac{1}{x}\right)y = 0, \qquad (103)$$

and that, for a fixed n, the functions named form the complete system of proper functions of the equation just written, when k runs through all non-negative integral values. Thus it follows that, for vanishing D, equation (41′) possesses the proper functions

$$u_k(\eta) = e^{-\frac{\eta}{2}}L_{n+k}^n(\eta) \qquad (49)$$

and the proper values

$$\frac{B}{\sqrt{-A}_+} = \frac{n+1}{2} + k \quad (k = 0,1,2,\ldots) \qquad (50)$$

– and no others! (See the mathematical appendix concerning the remarkable loss of the extended spectrum caused by the apparently inoffensive transformation (48); by this loss the development of the perturbation theory is made much easier.)

We have now to calculate the perturbation of the proper values (50) from the general theory of § 1, caused by including the D-term in (41′). The equation becomes self-adjoint if we multiply by η^{n+1}. The density function $\rho(x)$ of the general theory thus becomes η^n. As perturbation function $r(x)$ appears

$$-\frac{D}{(2\sqrt{-A}_+)^3}\eta^{n+2}. \qquad (51)$$

(We formally put the perturbation parameter $\lambda = 1$; if we desired, we could identify D or F with it.) Now formula (7′)

gives, for the perturbation of the kth proper value,

$$\epsilon_k = -\frac{D}{(2\sqrt{-A})^3_+} \frac{\int_0^\infty \eta^{n+2} e^{-\eta} [L_{n+k}^n(\eta)]^2 d\eta}{\int_0^\infty \eta^n e^{-\eta} [L_{n+k}^n(\eta)]^2 d\eta}. \qquad (52)$$

For the integral in the denominator, which merely provides for the normalisation, formula (115) of the appendix gives the value

$$\frac{[(n+k)!]^3}{k!}, \qquad (53)$$

while the integral in the numerator is evaluated in the same place, as

$$\frac{[(n+k)!]^3}{k!}(n^2 + 6nk + 6k^2 + 6k + 3n + 2). \qquad (54)$$

Consequently

$$\epsilon_k = -\frac{D}{(2\sqrt{-A})^3_+}(n^2 + 6nk + 6k^2 + 6k + 3n + 2). \qquad (55)$$

The condition for the kth perturbed proper value of equation (41') and therefore, naturally, also for the kth discrete proper value of the original equation (41) runs therefore

$$\frac{B}{\sqrt{-A}_+} = \frac{n+1}{2} + k + \epsilon_k \qquad (56)$$

(ϵ_k is retained meantime for brevity).

This result is applied twice, namely, to the last two equations of (38) by substituting the two systems (42) of values of the constants A, B, C, D; and it is to be observed that n is the *same* number in the two cases, while the two k-values are to be distinguished by the suffixes 1 and 2 , as above. First we have

$$
\begin{cases}
\dfrac{B_1}{\sqrt{-A}} = \dfrac{n+1}{2} + k_1 + \epsilon_{k_1} \\[2ex]
\dfrac{B_2}{\sqrt{-A}} = \dfrac{n+1}{2} + k_2 + \epsilon_{k_2}
\end{cases}
\tag{57}
$$

whence comes

$$
A = -\frac{(B_1 + B_2)^2}{(l + \epsilon_{k_1} + \epsilon_{k_2})^2}
\tag{58}
$$

(applying abbreviation (46) for the principal quantum number). In the approximation we are aiming at we may expand with respect to the small quantities ϵ_k and get

$$
A = -\frac{(B_1 + B_2)^2}{l^2}\left[1 - \frac{2}{l}(\epsilon_{k_1} + \epsilon_{k_2})\right].
\tag{59}
$$

Further, in the calculation of these small quantities, we may use the approximate value (45) for A in (55). We thus obtain, noticing the two D values, by (42),

$$
\begin{cases}
\epsilon_{k_1} = +\dfrac{Fh^4 l^3}{64\pi^4 m^2 e^5}(n^2 + 6nk_1 + 6k_1^2 + 6k_1 + 3n + 2). \\[2ex]
\epsilon_{k_2} = -\dfrac{Fh^4 l^3}{64\pi^4 m^2 e^5}(n^2 + 6nk_2 + 6k_2^2 + 6k_2 + 3n + 2).
\end{cases}
\tag{60}
$$

Addition gives, after an easy reduction,

$$
\epsilon_{k_1} + \epsilon_{k_2} = \frac{3Fh^4 l^4 (k_2 - k_1)}{32\pi^4 m^2 e^5}
\tag{61}
$$

If we substitute this, and the values of A, B_1 and B_2 from (42) in (59), we get, after reduction,

$$
E = -\frac{2\pi^2 m e^4}{h^2 l^2} - \frac{3}{8}\frac{h^2 F l (k_2 - k_1)}{\pi^2 m e}.
\tag{62}
$$

This is our provisional *conclusion*; it is the well-known formula of Epstein for the term values in the Stark effect of the hydrogen spectrum.

k_1 and k_2 correspond fully to the parabolic quantum numbers; they are capable of taking the value zero. Also the integer n, which has evidently to do with the *equatorial* quantum number, may from (40) take the value zero. However, from (46) the sum of these three numbers must still be increased by unity in order to yield the principal quantum number. Thus $(n + 1)$ and not n corresponds to the equatorial quantum number. The value zero for the *latter* is thus *automatically* excluded by wave mechanics, just as by Heisenberg's mechanics.[19] *There is simply no proper function,* i.e. no state of vibration, which corresponds to such a meridional orbit. This important and gratifying circumstance was already brought to light in Part I. in counting the constants, and also afterwards in § 2 of Part I. in connection with the azimuthal quantum number, through the non-existence of states of vibration corresponding to *pendulum orbits*; its full meaning, however, only fully dawned on me through the remarks of the two authors just quoted.

For later application, let us note the system of proper functions of equation (32) or (32′) in "zero approximation", which belongs to the proper values (62). It is obtained from statement (37), from conclusions (39) and (49), and from consideration of transformations (47) and (48) and of the approximate value (45) of A. For brevity, let us call a_0 the "radius of the first hydrogen orbit. Then we get

$$\frac{1}{2l\sqrt{-A}} = \frac{h^2}{4\pi^2 m e^2} = a_0 \qquad (63)$$

The proper functions (not yet normalised!) then read

$$\psi_{n k_1 k_2} = \lambda_1^{\frac{n}{2}} \lambda_2^{\frac{n}{2}} e^{-\frac{\lambda_1 + \lambda_2}{2 l a_0}} L_{n+k_1}^n \left(\frac{\lambda_1}{l a_0} \right) L_{n+k_2}^n \left(\frac{\lambda_2}{l a_0} \right) \frac{\sin}{\cos} n\phi. \qquad (64)$$

[19]W. Pauli, jun., *Ztschr. f. Phys.* 36, p. 336, 1926; N. Bohr, *Die Naturw.* 1, 1926.

They belong to the proper values (62), where l has the meaning (46). To each non-negative integral trio of values n, k_1, k_2 belong (on account of the double symbol $\begin{smallmatrix} \sin \\ \cos \end{smallmatrix}$) *two* proper functions or *one*, according as $n > 0$ or $n = 0$.

§ 4. Attempt to calculate the Intensities and Polarisations of the Stark Effect Patterns

I have lately shown[20] that from the proper functions we can calculate by differentiation and quadrature the elements of the *matrices*, which are allied in Heisenbergs mechanics to functions of the generalised position- and momentum-coordinates. For example, for the (rr')th element of the matrix, which according to Heisenberg belongs to the generalised coordinate q itself, we find

$$\begin{cases} q^{rr'} = \int q\rho(x)\psi_r(x)\psi_{r'}(x)dx \\ \cdot \left\{ \int \rho(x)[\psi_r(x)]^2 dx \cdot \int \rho(x)[\psi_{r'}(x)]^2 dx \right\}^{-\frac{1}{2}} . \end{cases} \quad (65)$$

Here, for our case, the separate indices *each* deputise for a *trio* of indices n, k_1, k_2, and further, x represents the three coordinates r, θ, ϕ. $\rho(x)$ is the density function; in our case the quantity (34). (We may compare the self-adjoint equation (32') with the general form (2)). The "denominator" $(\ldots)^{-\frac{1}{2}}$ in (65) must be put in because our system (64) of functions is not yet normalised.

According to Heisenberg,[21] now, if q means a rectangular Cartesian coordinate, then the square of the matrix element (65) is to be a measure of the "probability of transition" from

---

[20]Preceding paper of this collection.

[21]W. Heisenberg, *Ztschr. f. Phys.* **33**, p. 879, 1925; M. Born and P. Jordan, *Ztschr. f. Phys.* **34**, pp. 867, 886, 1925.

the rth state to the r'th, or, more accurately, a measure of the intensity of that part of the radiation, bound up with this transition, which is polarised in the q-direction. Starting from this, I have shown in the above paper that if we make certain simple assumptions as to the electrodynamical meaning of ψ, the "mechanical field scalar", then the matrix element in question is susceptible of a very simple physical interpretation in wave mechanics, namely, *actually*: component of the amplitude of the periodically oscillating electric moment of the atom. The word *component* is to be taken in a double sense: (1) component in the q-direction, i.e. in the spatial direction in question, and (2) only the part of this spatial component which changes in a time-sinusoidal manner with exactly the frequency of the emitted light, $|E_r - E_{r'}|/h$. (It is a question then of a kind of Fourier analysis: not in harmonic frequencies, but in the actual frequencies of emission.) However, the idea of wave mechanics is not that of a sudden transition from one state of vibration to another, but according to it, the partial moment concerned – as I will briefly name it – arises from the *simultaneous existence* of the two proper vibrations, and lasts just as long as both are excited together.

Moreover, the above assertion that the $q^{rr'}$'s are proportional to the partial moments is more accurately phrased thus. The ratio of, e.g., $q^{rr'}$ to $q^{rr''}$ is equal to the ratio of the partial moments which arise when the proper function ψ_r and the proper functions $\psi_{r'}$ and $\psi_{r''}$ are stimulated, the first *with any strength whatever* and the last two with strengths *equal to one another* – i.e. corresponding to normalisation. To calculate the ratio of the *intensities*, the q-quotient must first be squared and then multiplied by the ratio of the fourth powers of the emission frequencies. The latter, however, has no part in the intensity ratio of the Stark effect components, for there we only compare intensities of lines which have practically the same frequency.

The known *selection* and *polarisation rules* for Stark ef-

134

fect components can be obtained, almost without calculation, from the integrals in the numerator of (65) and from the form of the proper functions in (64). They follow from the vanishing or non-vanishing of the integral with respect to ϕ. We obtain the components whose electric vector vibrates *parallel* to the field, i.e. to the z-direction, by replacing the q in (65) by z from (33). The expression for z, i.e. $\frac{1}{2}(\lambda_1 - \lambda_2)$, does *not* contain the azimuth ϕ. Thus we see at once from (64) that a non-vanishing result after integration with respect to ϕ can only arise if we combine proper functions whose n's are *equal*, and thus whose equatorial quantum numbers are equal, being in fact equal to $n+1$. For the components which vibrate *perpendicular* to the field, we must put q equal to x or equal to y (cf. equation (33)). Here $\cos\phi$ or $\sin\phi$ enters, and we see almost as easily as before, that the n-values of the two combined proper functions must differ exactly by unity, if the integration with respect to ϕ is to yield a non-vanishing result. Hence the known selection and polarisation rules are proved. Further, it should be recalled again that we do not require to exclude any n-value after additional reflection, as was necessary in the older theory in order to agree with experience. Our n is smaller by 1 than the equatorial quantum number, and right from the beginning cannot take negative values (quite the same state of affairs exists, we know, in Heisenberg's theory).[22]

The numerical evaluation of the integrals with respect to λ_1 and λ_2 which appear in (65) is exceptionally tedious, especially for those of the numerator. The same apparatus for calculating comes into play as served already in the evaluation of (52), only the matter is somewhat more detailed because the two (generalised) Laguerre polynomials, whose product is to be integrated, have not the same argument. By good luck, in the *Balmer lines*, which interest us principally, one of the

---

[22] W. Pauli, jun., *Ztschr. f. Physik*, 36, p. 336, 1926.

two polynomials L^n_{n+k}, namely that relating to the doubly quantised state, is either a constant or is a linear function of its argument. The method of calculation is described more fully in the mathematical appendix. The following tables and diagrams give the results for the first four Balmer lines, in comparison with the known measurements and estimates of intensity, made by Stark[23] for a field strength of about 100,000 volts per centimetre. The first column indicates the state of polarisation, the second gives the combination of the terms in the usual manner of description, i.e. in *our* symbols: of the two trios of numbers $(k_1, k_2, n+1)$ the *first* trio refers to the higher quantised state and the *second* to the doubly quantised state. The third column, with the heading Δ, gives the term decomposition in multiples of $3h^2F/8\pi^2 me$, (see equation (62)). The next column gives the intensities observed by Stark, and 0 there signifies not observed. The question mark was put by Stark at such lines as clash either with irrelevant lines or with possible "ghosts" and thus cannot be guaranteed. On account of the unequal weakening of the two states of polarisation in the spectrograph, according to Stark his results for the $\parallel$ and for the $\perp$ components of vibration are not directly comparable with one another. Finally, the last column gives the results of our calculation in *relative numbers*, which are comparable for the collective components ($\parallel$ and $\perp$) of *one* line, e.g. of H_α, but not for those of H_α with H_β, etc. These relative numbers are reduced to their *smallest integral* values, i.e. the numbers in each of the four tables are *prime* to each other.

[23] J. Stark, *Ann. d. Phys.* 48, p. 193, 1915.

TABLES

INTENSITIES IN THE STARK EFFECT OF THE BALMER LINES

TABLE 1

H_α

Polarisation.	Combination.	Δ	Observed Intensity.	Calculated Intensity.
‖	(111) (011)	2	1	729
	(102) (002)	3	1·1	2304
	(201) (101)	4	1·2	1681
	(201) (011)	8	0	1
				Sum : 4715
⊥	(003) (002)	0	} 2·6 {	4608
	(111) (002)	0		882
	(102) (101)	1	1	1936
	(102) (011)	5	0	16
	(201) (002)	6	0	18
				Sum * : 4715

* Undisplaced components halved.

TABLE 2

H_β

Polarisation.	Combination.	Δ	Observed Intensity.	Calculated Intensity.
‖	(112) (002)	0	1·4	0
	(211) (101)	2	1·2	9
	—	(4)	1	0
	(211) (011)	6	4·8	81
	(202) (002)	8	9·1	384
	(301) (101)	10	11·5	361
	—	(12)	1	0
	(301) (011)	14	0	1
				Sum : 836
⊥	—	(0)	1·4	0
	(112) (011)	2	3·3	72
	(103) (002)	4	} 12·6 {	384
	(211) (002)	4		72
	(202) (101)	6	9·7	294
	—	(8)	1·3	0
	(202) (011)	10	1·1 ?	6
	(301) (002)	12	1 ?	8
				Sum : 836

INTENSITIES IN THE STARK EFFECT OF THE BALMER LINES

TABLE 3

H_γ

Polarisation.	Combination.	Δ	Observed Intensity.	Calculated Intensity.
‖	(221) (011)	2	1·6	15 625
	(212) (002)	5	1·5	19 200
	(311) (101)	8	1	1 521
	(311) (011)	12	2·0	16 641
	(302) (002)	15	7·2	115 200
	(401) (101)	18	10·8	131 769
	(401) (011)	22	1 ?	729
				Sum : 300 685
⊥	(113) (002)	0	} 7·2	{ 115 200
	(221) (002)	0		26 450
	(212) (101)	3	3·2	46 128
	(212) (011)	7	1·2	5 808
	(203) (002)	10	} 4·3	{ 76 800
	(311) (002)	10		11 250
	(302) (101)	13	6·1	83 232
	(302) (011)	17	1·1	2 592
	(401) (002)	20	1	4 050
				Sum : * 300 685

* Undisplaced components halved.

TABLE 4

H_δ

Polarisation.	Combination.	Δ	Observed Intensity.	Calculated Intensity.
‖	(222) (002)	0	0	0
	(321) (101)	4	1	8
	(321) (011)	8	1·2	32
	(312) (002)	12	1·5	72
	(411) (101)	16	1·2	18
	(411) (011)	20	1·1	18
	(402) (002)	24	2·8	180
	(501) (101)	28	7·2	242
	(501) (011)	32	1 ?	2
				Sum: 572
⊥	(222) (011)	2	1·3	36
	(213) (002)	6	} 3·2	{ 162
	(321) (002)	6		36
	(312) (101)	10	2·1	98
	(312) (011)	14	1	2
	(303) (002)	18	} 2·0	{ 90
	(411) (002)	18		9
	(402) (101)	22	2·4	125
	(402) (011)	26	1·3	5
	(501) (002)	30	1 ?	9
				Sum : 572

138

In the *diagrams* it is to be noticed that, on account of the huge differences in the theoretical intensities, some theoretical intensities cannot be truly represented to scale, as they are much too small. These are indicated by *small circles*.

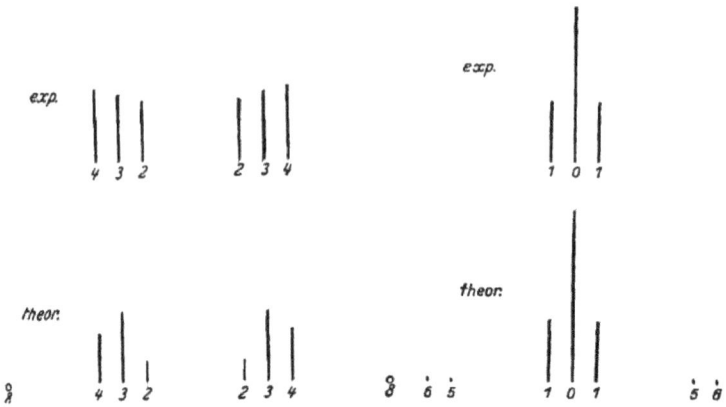

FIG. 1.—H_a ‖-components.

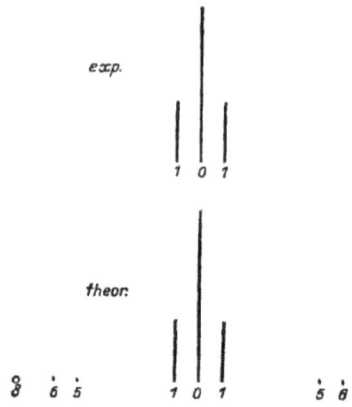

FIG. 2.—H_a ⊥-components

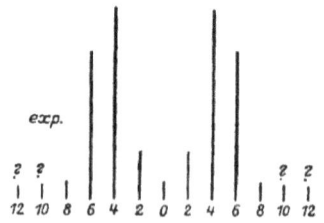

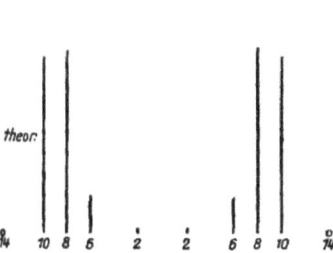

FIG. 3.—H_β ‖-components.

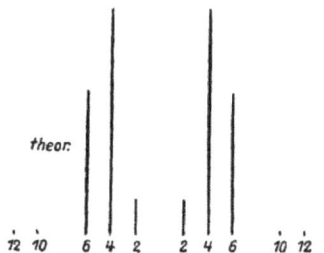

FIG. 4.—H_β ⊥-components.

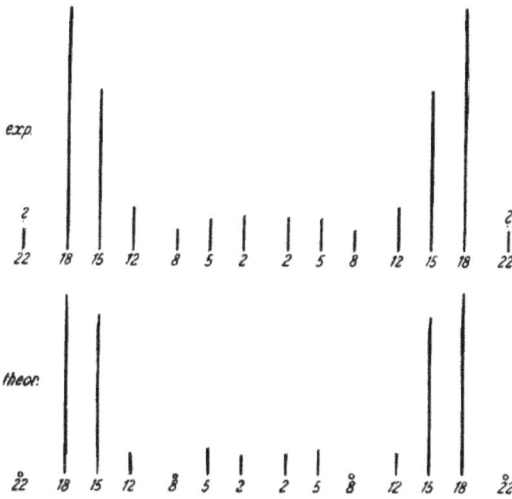

Fig. 5.—H_γ ‖-components.

A consideration of the diagrams shows that the agreement is tolerably good for almost all the strong components, and taken all over it is somewhat better than for the values deduced from correspondence considerations.[24] Thus, for example, it is removed one of the most serious contradictions which arose, in that the correspondence principle gave the ratio of the intensities of the two strong ⊥-components of H_β, for $\Delta = 4$ and 6, inversely and indeed very much out, in fact as almost 1 : 2, while experiment requires about 5 : 4. A similar thing occurs with the mean ($\Delta = 0$) ⊥-components of H_γ, which decidedly preponderate experimentally, but are given as *far* too weak by the correspondence principle. In *our* diagrams also, it is admitted that such "reciprocities" between the intensity ratios of intense components demanded by theory and by experiment are not entirely wanting. The theoretically most intense ‖-component ($\Delta = 3$) of H_α is furthest out; by experiment, it should lie *between* its neighbours in intensity. And the two strongest ‖-components of H_β and

[24]H. A. Kramers, *Dänische Akademie* (8), iii. 3, p. 333 *et seq.*, 1919.

two $\perp$-components ($\Delta = 10, 13$) of H_γ are given "reciprocally" by the theory. Of course, in both cases the intensity ratios, both experimentally and theoretically, are pretty near unity.

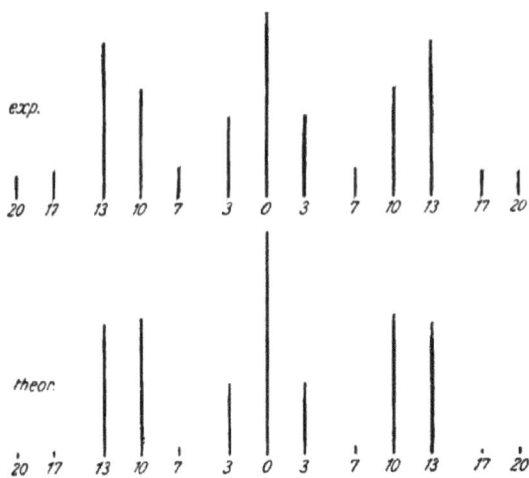

FIG. 6.—H_γ $\perp$-components.

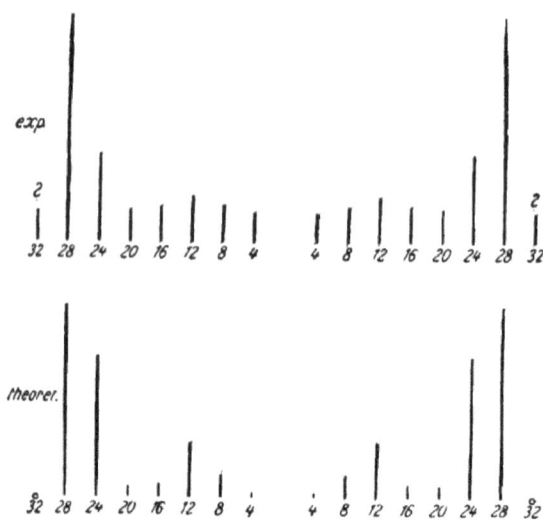

FIG. 7.—H_δ ||-components.

Passing now to the weaker components, we notice first that the contradiction which exists for some weak observed components of H_β to the selection and polarisation rules, of course still remains in the new theory, since the latter gives these rules in conformity with the older theory. However, components which are extremely weak theoretically are for the most part *un*observed, or the observations are *questionable*. The strength *ratios* of weaker components to one another or to stronger ones are *almost never* given even approximately correctly; cf. especially H_γ y and H_δ. Such serious mistakes in the experimental determination of the blackening are of course out of the question.

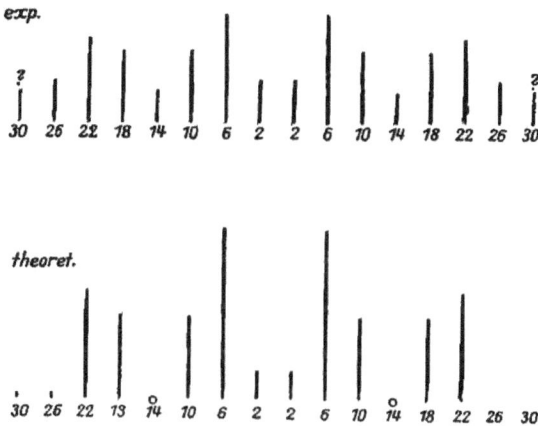

FIG. 8.—H_δ ⊥-components.

Considering all this, we might feel inclined to be very sceptical of the thesis that the integrals (65) or their squares are measures of intensity. I am far from wishing to represent this thesis as irrefutable. There are still many alterations conceivable, and these may, perhaps, be necessitated by internal reasons when the theory is further extended. Yet the following should be remembered. The whole calculation has been performed with the *unperturbed* proper functions, or more precisely, with the *zero* approximation to the perturbed ones (cf. above § 2). It, therefore, represents an approximation for a *vanishing* field strength! However, just for the weak or almost vanishing components we should expect theoretically a fairly powerful growth with increasing field strength, for the following reason. According to the view of wave mechanics, as explained at the beginning of this section, the integrals (65) represent the amplitudes of the electrical partial moments, which are produced by the distribution of charges which flow round about the nucleus within the atom's domain. When for a line component we get as a zero approximation very weak or even vanishing intensity, this is not caused in any way by the fact that to the simultaneous existence of the two proper vibrations corresponds only an insignificant motion of electricity, or even none at all. The vibrating mass of electricity – if this vague expression is allowed – may be represented as the same in all components, on the ground of *normalisation*. Rather is the reason for the low line intensity to be found in a high degree of *symmetry* in the motion of the electricity, through which only a small, or even no, dipole moment arises (on the contrary, e.g., only a four-pole moment). Therefore it is to be expected that the *vanishing* of a line component in presence of perturbations of any kind is a relatively *unstable* condition, since the symmetry is probably destroyed by the perturbation. And thus it may be expected that weak or vanishing components gain quickly in intensity with increasing field strength.

This has now actually been observed, and the intensity ratios, indeed, alter quite considerably with field strength, for strengths of about 10,000 gauss and upwards; and, if I understand aright, in the way[25] shown by the present general discussion. Certain information on the question whether this really explains these discrepancies could of course only be got from a continuation of the calculation to the next approxima-

---

[25] J. Stark, *Ann. d. Phys.* 43, p. 1001 *et seq.*, 1914.

tion, but this is *very* troublesome and complicated.

The present considerations are of course nothing but the "translation" into the language of the new theory of very well-known considerations which Bohr[26] has brought forward in connection with calculation of line intensities by means of the principle of correspondence.

The theoretical intensities given in the tables satisfy a fundamental requirement, which is set up not only by intuition but also by experiment,[27] viz., the sum of the intensities of the ‖-components is equal to that of the ⊥-components. (Before adding, *undisplaced* components must be *halved* – as a compensation for the *duplication* of all the others, which occur on both sides.) This makes a very welcome "control" for the arithmetic.

It is also of interest to compare the *total intensities* of the four lines by using the four "sums" given in the tables. For this purpose I take back from my numerical calculations the four factors, which were omitted in order to represent the intensity ratios within each of the four line groups by the smallest integers possible, and multiply by them. Further, I multiply each of these four products by the *fourth* power of the appropriate emission frequency. Thus I obtain the following four numbers:

$$\text{for } H_\alpha \ldots \quad \frac{2^6 \cdot 23 \cdot 41}{3^2 \cdot 5^9} = 0.003433\ldots$$

$$\text{for } H_\beta \ldots \quad \frac{4 \cdot 11 \cdot 19}{3^{12}} = 0.001573\ldots$$

$$\text{for } H_\gamma \ldots \quad \frac{2^6 \cdot 3^6 \cdot 11^2 \cdot 71}{5 \cdot 7^{13}} = 0.0008312\ldots$$

$$\text{for } H_\delta \ldots \quad \frac{11 \cdot 13}{2^{15}3^2} = 0.0004849\ldots$$

I give these numbers with still greater reserve than the

[26]N. Bohr, *Dänische Akademie* (8), iv. I. 1, p. 35, 1918.
[27]J. Stark, *Ann. d. Phys.* 43, p. 1004, 1914.

former ones because I am not sure, theoretically, about the *fourth* power of the frequency. Investigations[28] which I have lately published seem to call, perhaps, for the *sixth*. The above method of calculation corresponds exactly to the assumptions of Born, Jordan, and Heisenberg.[29] Fig. 9 represents the results diagrammatically.

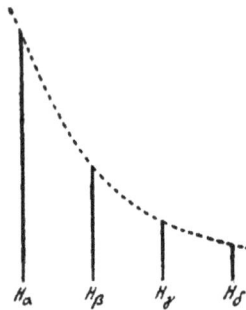

FIG. 9.—Total Intensities.

Actual measured intensities of emission lines, which are known to depend greatly on the conditions of excitation, naturally cannot here be used in a comparison with experience. From his researches[30] on dispersion and magneto-rotation in the neighbourhood of H_α and H_β, R. Ladenburg has, with F. Reiche,[31] calculated the value 4·5 (limits 3 and 6) for the ratio of the so-called "electronic numbers" of these two lines. If I assume that the above numbers may be taken as proportional

---

[28] Equation (38) at end of previous paper of this collection. The *fourth* allows for the fact that for the radiation it is a question of the square of the *acceleration* and not of the electric moment itself. In this equation (38) occurs explicitly another factor $(E_k - E_m)/h$. This is occasioned by the appearance of $\frac{\partial}{\partial t}$ in statement (36).
Addition at proof correction: Now I recognise this $\frac{\partial}{\partial t}$ to be incorrect, though I hoped it would make the later relativistic generalisation easier. Statement (36), *loc. cit.*, is to be replaced by $\psi\bar\psi$. The above doubts about the *fourth* power are therefore dissolved.

[29] Cf. M. Born and P. Jordan, *Ztschr. f. Phys.* 34, p. 887, 1925.

[30] R. Ladenburg, *Ann. d. Phys.* (4), 38, p. 249, 1912.

[31] R. Ladenburg and F. Reiche, *Die Naturwissenschaften*, 1923, p. 584.

to Ladenburg's[32] expression,

$$\sum \frac{g_k}{g_i} a_{ki} \nu_0,$$

then they may be reduced to (relative) "electronic numbers" by division by ν_0, i.e. by

$$\left(\frac{5}{36}\right)^3, \ \left(\frac{3}{16}\right)^3, \ \left(\frac{21}{100}\right)^3, \ \text{and} \ \left(\frac{2}{9}\right)^3, \ \text{respectively.}$$

Hence we obtain the four numbers,

$$1.281, \ 0.2386, \ 0.08975, \ 0.04418.$$

The ratio of the first to the second is 5.37, which agrees sufficiently with Ladenburg's value.

§ 5. Treatment of the Stark Effect by the Method which corresponds to that of Bohr

Mainly to give an *example* of the general theory of § 2, I wish to outline *that* treatment of the proper value problem of equation (32), which must have been adopted, if we had *not* noticed that the perturbed equation is also exactly "separable" in parabolic coordinates. We therefore now keep to the polar coordinates r, θ, ϕ, and thus replace z by $r\cos\theta$. We also introduce a new variable η for r by the transformation

$$2r\sqrt{-\frac{8\pi^2 m E}{h^2}} = \eta, \tag{66}$$

(which is closely akin to transformation (48) for the parabolic coordinate ξ). For one of the unperturbed proper values (45),

---

[32] Cf. Ladenburg-Reiche, *loc. cit.*, the first formula in the second column, p. 584. The factor ν_0 in the *above* expression comes from the fact that the "transition probability" a_{ki} is still to be multiplied by the "energy quantum" to give the intensity of the radiation.

146

we get from (66)

$$\eta = \frac{2r}{lA_0},\qquad(66')$$

where a_0 is the same constant as in (63). ("Radius of the innermost hydrogen orbit.") If we introduce this and the unperturbed value (45) into the equation (32), which is to be treated, then we obtain

$$\nabla'^2\psi + \left(-\frac{1}{4} - g\eta\,\cos\theta + \frac{l}{\eta}\right)\psi = 0,\qquad(67)$$

where for brevity

$$g = \frac{a_0^2 F l^3}{4e}.\qquad(68)$$

The dash on the Laplacian operator is merely to signify that in it the letter η is to be written for the radius vector.

In equation (67) we *conceive l to be the proper value*, and the term in g to be the perturbing term. The fact that the perturbing term *contains* the proper value need not trouble us in the first approximation. If we neglect the perturbing term, the equation has as proper values the natural numbers

$$l = 1, 2, 3, 4\ldots\qquad(69)$$

and no others. (The extended spectrum is again cut out by the artifice (66), which would be valuable for closer approximations.) The allied *proper functions* (not yet normalised) are

$$\psi_{lnm} = P_n^m(\cos\theta)\,\frac{\sin}{\cos}\,(m\phi)\cdot\eta^n e^{-\frac{\eta}{2}} L_{n+l}^{2n+1}(\eta).\qquad(70)$$

Here P_n^m signifies the mth "associated" Legendre function of the nth order, and L_{n+l}^{2n+1} is the $(2n+l)$th derivative of the $(n+1)$th Laguerre polynomial.[33] So we must have

$$n < l,$$

---

[33]I lately gave the proper functions (70) (see Part I.), but without noticing their connection with the Laguerre polynomials. For the proof of the above representation, see the Mathematical Appendix, section I.

otherwise L_{n+l}^{2n+1} would vanish, because the number of differentiations would be greater than the degree. With reference to this, the numbering of the spherical surface harmonics shows that l is an l^2-fold proper value of the unperturbed equation. We now investigate the *splitting up* of a *definite* value of l, supposed fixed in what follows, due to the addition of the perturbing term.

To do this we have, *in the first place*, to normalise our proper functions (70), according to § 2. From an uninteresting calculation, which is easily performed with the aid of the formulae in the appendix,[34] we get as the normalising factor

$$\frac{1}{\sqrt{\pi}}\sqrt{\frac{2n+1}{2}}\sqrt{\frac{(n-m)!}{(n+m)!}}\sqrt{\frac{(l-n-1)!}{[(n+l)!]^3}}, \qquad (71)$$

if $m \neq 0$, but, for $m = 0$, $\frac{1}{\sqrt{2}}$ times this value. *Secondly,* we have to calculate the symmetrical matrix of constants ϵ_{im}, according to (22). The r there is to be identified with our perturbing function $-g\eta^3\cos\theta\sin\theta$, and the proper functions, there called u_{ki}, are to be identified with our functions (70). The fixed suffix k, which characterises the proper value, corresponds to the *first* suffix l of ψ_{lnm}, and the *other* suffix i of u_{ki} corresponds now to the pair of suffixes n, m in ψ_{lnm}. The matrix (22) of constants forms in our case a square of l^2 rows and l^2 columns. The quadratures are easily carried out by the formulae of the appendix and yield the following results. Only those elements of the matrix are different from zero, for which the two proper functions ψ_{lnm}, $\psi_{ln'm'}$, to be combined, satisfy the following conditions simultaneously:

1. The *upper indices* of the "associated Legendre functions" must agree, i.e. $m = m'$.

---

[34]It is to be noticed that the *density function*, generally denoted by $\rho(x)$, reads as $\eta\sin\theta$ in equation (67), because the equation must be multiplied by $\eta^2\sin\theta$, in order to acquire self-adjoint form.

2. The *orders* of the two Legendre functions must differ exactly by unity, i.e. $|n - n'| = 1$.

3. To each trio of indices lnm, if $m \neq 0$, there belong, according to (70), *two* Legendre functions, and thus also two proper functions ψ_{lnm}, which only differ from each other in that one contains a factor $\cos m\phi$ and the other $\sin m\phi$. The third condition reads: we may only combine sine with sine, or cosine with cosine, and not sine with cosine.

The remaining non-vanishing elements of the desired matrix would have to be characterised from the beginning by *two* index-*pairs* (n, m) and $(n + 1, m)$. (We renounce any idea of showing the fixed index l explicitly.) Since the matrix is symmetrical, *one* index pair (n, m) is sufficient, if we stipulate that the first index, i.e. n, shall mean the *greater* of the two orders n, n', in every case.

Then the calculation gives

$$\epsilon_{nm} = -6lg\sqrt{\frac{(l^2 - n^2)(n^2 - m^2)}{4n^2 - 1}}. \tag{72}$$

We have now to form the determinant (22) out of these elements. It is advantageous to *arrange* its rows *as well as* its columns on the following principle. (To fix our ideas, let us speak of the columns, and therefore of the index-pair characterising the *first* of the two Legendre functions.) Thus: first come all terms with $m = 0$, then all with $m = 1$, then all with $m = 2$, etc., and finally, all terms with $m = l - 1$, which last is the greatest value that m (like n) can take. *Inside* each of these groups, let us arrange the terms thus: first, all terms with $\cos m\phi$, and then all with $\sin m\phi$. Within these "half groups" let us arrange them in order of increasing n, which runs through the values $m, m + 1, m + 2, \ldots l - 1$, i.e. $(l - m)$ values in all.

If we carry this out, we find that the non-vanishing elements (72) are exclusively confined to the two secondary diagonals, which lie immediately alongside the principal diagonal. On the latter are the proper value perturbations which are to be found, but taken negatively, while everywhere else are zeros. Further, the two secondary diagonals are interrupted by zeros at *those* places, where they break through the *boundaries* between the so-called "half-groups", in very convenient fashion. Hence the whole determinant *breaks up* into a product of just so many smaller determinants as there are "half-groups" present, viz. $(2l - 1)$. It will be sufficient if we consider one of them. We write it here, denoting the desired perturbation of the proper value by ϵ (without suffix):

$$
\begin{vmatrix}
-\epsilon & \epsilon_{m+1,m} & 0 & 0 & \cdots & 0 \\
\epsilon_{m+1,m} & -\epsilon & \epsilon_{m+2,m} & 0 & \cdots & 0 \\
0 & \epsilon_{m+2,m} & -\epsilon & \epsilon_{m+3,m} & \cdots & 0 \\
0 & 0 & \epsilon_{m+3,m} & -\epsilon & \cdots & 0 \\
\cdots & \cdots & \cdots & \cdots & \cdots & \cdots \\
\cdots & \cdots & \cdots & \cdots & \cdots & \cdots \\
0 & 0 & 0 & 0 & \epsilon_{l-1,m} & -\epsilon
\end{vmatrix}
\quad (73)
$$

If we divide each term here by the common factor $6lg$ of the ϵ_{nm}'s (cf. (72)), and for the moment regard as the unknown

$$
k^* = \pm -\frac{\epsilon}{6lg}, \quad (74)
$$

the above equation of the $(l - m)$th degree has the roots

$$
k^* = \pm(l - m - 1),\ \pm(l - m - 3),\ \pm(l - m - 5)\ldots \quad (75)
$$

where the series stops with ± 1 or 0 (inclusive) according as the *degree* $l - m$ is *even* or *odd*. The proof of this is unfortunately *not* to be found in the appendix, as I have not been successful in obtaining it.

If we form the series (75) for each of the values $m = 0, 1, 2, \ldots (l - 1)$, then we have in the numbers

$$
\epsilon = -6lgk^* \quad (76)
$$

the complete set of *perturbations of the principal quantum number l*. In order to find the perturbed *proper* values E (term-levels) of the equation (32), we have only to substitute (76) in

$$E = -\frac{2\pi^2 me^4}{h^2(l+\epsilon)^2} \tag{77}$$

taking into account the signification of the abbreviations g (see (68)) and a_0 (see (63)).

After reducing this gives

$$E = -\frac{2\pi^2 me^4}{h^2 l^2} - \frac{3}{8}\frac{h^2 Flk^*}{\pi^2 me}. \tag{78}$$

Comparison with (62) shows that k^* is the *difference* $k_2 - k_1$ of the parabolic quantum numbers. From (75), bearing in mind the range of values of m referred to above, we see that k^* may also take the same values as the difference just mentioned, viz. $0, 1, 2, \ldots (l-1)$. Also, if we take the trouble to work it out, we will find for the *multiplicity*, in which k^* and the difference $k_2 - k_1$ appear, the same value, viz. $l - |k^*|$.

We have thus obtained the proper value perturbations of the first order also from the general theory. The next step would be the solution of the system (21') of linear equations of the general theory for the κ-quantities. These would then yield, according to (18) (provisionally putting $\lambda = 0$), the perturbed proper functions of zero order; this is nothing more than a representation of the proper functions (64) as linear forms of the proper functions (70). In our case the solution of (21') would naturally be anything but unique, on account of the considerable multiplicity of the roots ϵ. The solution is made much simpler if we notice that the equations break up into *just as many* groups, viz. (2l−1), or, retaining the former expression, *half-groups*, with completely separated variables, as the determinant investigated above contains factors like (73); and if we further notice that it is allowable, after we have chosen a definite ϵ-value, to regard only the variables

κ of a *single* half-group as different from zero, of that half-group, in fact, for which the determinant (73) vanishes for the chosen ϵ-value. The definition of this half-group of variables is then *unique*.

But our object, viz. to illustrate the general method of § 2 by an example, has been sufficiently attained. Since the continuation of the calculation is of no special physical interest, I have not troubled to bring the determinantal quotients, which we immediately obtain for the coefficients κ, into a clearer form, or to work out the transformation to principal axes in any other way.

On the whole, we must admit that in the present case the method of secular perturbations (§5) is considerably more troublesome than the direct application of a system of separation (§ 3). I believe that this may also be true in other cases. In ordinary mechanics it is, as we know, usually quite the reverse.

III.–MATHEMATICAL APPENDIX

Prefatory Note: – It is not intended to supply in uninterrupted detail all the calculations omitted from the text. Without that, the present paper has already become too long. In general, only those methods of calculation will be briefly described which another might utilise with advantage in similar work, if something better does not occur to him – as it may easily do.

§ 1. The Generalised Laguerre Polynomials and Orthogonal Functions

The kth Laguerre polynomial $L_k(x)$ satisfies the differential equation[35]

$$xy'' + (1 - x)y' + ky = 0. \tag{101}$$

[35] Courant-Hilbert, chap. ii. § 11, 5, p. 78, equation (72).

If we first replace k by $n + k$, and then differentiate n times, we find that the nth derivative of the $(n + k)$th Laguerre polynomial, which we will always denote by L_{n+k}^n, satisfies the equation

$$xy'' + (n + 1 - x)y' + ky = 0. \tag{103}$$

Moreover, by an easy transformation, we find that for $e^{-\frac{x}{2}} L_{n+k}^n(x)$ the following equation holds,

$$y'' + \frac{n+1}{x}y' + \left(-\frac{1}{4} + \left(k + \frac{n+1}{2} \right) \frac{1}{x} \right) y = 0. \tag{103}$$

This found an application in equation $(41')$ of § 3. The allied generalised Laguerre orthogonal functions are

$$x^{\frac{n}{2}} e^{-\frac{x}{2}} L_{n+k}^n(x). \tag{104}$$

Their equation, it may be remarked in passing, is

$$y'' + \frac{1}{x}y' + \left(-\frac{1}{4} + \left(k + \frac{n+1}{2} \right) \frac{1}{x} - \frac{n^2}{4x^2} \right) y = 0. \tag{105}$$

Let us turn to equation (103), and consider there that n is a fixed (real) integer, and k is the proper value parameter. Then, according to what has been said, in the domain $x \geq 0$, at any rate, the equation has the proper functions,

$$e^{-\frac{x}{2}} L_{n+k}^n(x), \tag{106}$$

belonging to the proper values,

$$k = 0, 1, 2, 3, \ldots \tag{107}$$

In the text it is maintained that it has no further values, and, above all, that it possesses no continuous spectrum. This seems paradoxical, for the equation

$$\frac{d^2 y}{d\xi^2} + \frac{n+1}{\xi} \frac{dy}{d\xi} + \left(-\frac{1}{(2k+n+1)^2} + \frac{1}{\xi} \right) y = 0 \tag{108}$$

into which (103) is transformed by the substitution

$$\xi = \left(k + \frac{n+1}{2} \right) x \tag{109}$$

does possess a continuous spectrum, if in it we regard

$$E = -\frac{1}{(2k + n + 1)^2} \tag{110}$$

as proper value parameter, viz. all positive values of E are proper values (cf. Part I., analysis of equation (7)). The reason why *no* proper values k of (103) can correspond to these positive E-values is that by (110) the k-values in question would be complex, and this is impossible, according to general theorems.[36] Each *real* proper value of (103), by (110), gives rise to a *negative* proper value of (108). Moreover, we know (cf. Part I.) that (108) possesses absolutely no negative proper values other than those that arise, as in (110), from the series (107). There thus remains only the one possibility, that in the series (107) certain negative k-values are lacking, which appear on solving (110) for k, on account of the double-valuedness when extracting the root. But this also is impossible, because the k-values in question turn out to be algebraically less than $-\frac{n+1}{2}$ and thus, from general theorems,[37] cannot be proper values of equation (103). The series of values (107) is thus complete. Q.E.D.

The above supplements the proof that the functions (70) are the proper functions of (67) (with the perturbing term suppressed), allied to the proper values (69). We have only to write the solutions of (67) as a product of a function of θ, ϕ and a function of η. The equation in η can readily be brought to the form of (105), the only difference being that our present n is there always an odd number, namely, the $(2n + 1)$ which is to be found there.

---

[36] Courant-Hilbert, chap. iii. § 4, 2, p. 115.
[37] Courant-Hilbert, chap. v. § 5, 1, p. 240.

154

§ 2. Definite Integrals of Products of Two Laguerre Orthogonal Functions

The Laguerre polynomials can all be obtained, in the following manner, as coefficients of the powers of the auxiliary variable t, in the expansion in a series of a so-called "generating function"[38]

$$\sum_{k=0}^{\infty} L_k(x)\frac{t^k}{k!} = \frac{e^{-\frac{xt}{1-t}}}{1-t}. \tag{111}$$

If we replace k by $n+k$ and then differentiate n times with respect to x, we obtain the generating function of our generalised polynomials,

$$\sum_{k=0}^{\infty} L_{n+k}^n(x)\frac{t^k}{(n+k)!} = (-1)^n\frac{e^{-\frac{xt}{1-t}}}{(1-t)^{n+1}}. \tag{112}$$

In order to evaluate with its help integrals such as appeared for the first time in the text in expression (52), or, more generally, such as were necessary in §4 for the calculation of (65), and also in §5, we proceed as follows. We write (112) over again, providing both the fixed index n and the varying index k with a dash, and replacing the undefined t by s. These two equations are then multiplied together, i.e. left side by left side, and right side by right. Then we multiply further by

$$x^p e^{-x} \tag{113}$$

and integrate with respect to x from 0 to ∞. p is to be a positive integer – this being sufficient for our purpose. The integration is practicable by elementary methods on the right-

---

[38]Courant-Hilbert, chap. ii. § 11, 5, p. 78, equation (68).

hand side, and we get

$$\sum_{k=0}^{\infty} \sum_{k'=0}^{\infty} \frac{t^k s^{k'}}{(n+k)!(n'+k')!} \int_0^{\infty} x^p e^{-x} L_{n+k}^n(x) L_{n'+k'}^{n'}(x) dx$$

$$= (-1)^{n+n'} p! \frac{(1-t)^{p-n}(1-s)^{p-n'}}{(1-ts)^{p+1}}.$$

$$(114)$$

We have now, on the left, the desired integrals like pearls on a string, and we merely detach the one we happen to need by searching on the right for the coefficient of $t^k s^{k'}$. This coefficient is always a simple sum, and, in fact, in the cases occurring in the text, always a finite sum with very few terms (up to three). In general, we have

$$\begin{cases} \int_0^{\infty} x^p e^{-x} L_{n+k}^n(x) L_{n'+k'}^{n'}(x) dx = p!\,(n+k)!\,(n'+k')! \\[2mm] \cdot \sum_{\tau=0}^{\leq k, k'} (-1)^{n+n'+k+k'+\tau} \binom{p-n}{k-\tau}\binom{p-n'}{k'-\tau}\binom{-p-1}{\tau}. \end{cases}$$

$$(115)$$

The sum stops after the smaller of the two numbers k, k'. It often, in actual fact, begins at a positive value of τ, as binomial coefficients, whose lower number is greater than the upper, vanish. For example, in the integral in the denominator of (52), we put $p = n = n'$ and $k' = k$. Then τ can take only the *one* value k, and we can establish statement (53) of the text. In the integral of the numerator in (52), only p has another value, namely $p = n + 2$. τ now takes the values $k-2, k-1$, and k, and after an easy reduction we get formula (54) of the text. In the very same way the integrals appearing in §5 are evaluated by Laguerre polynomials.

We can now, therefore, regard integrals of the type of (115) as known, and we have only to concern ourselves with those occurring in § 4 in the calculation of intensities (cf.

156

expression (65) and functions (64) which have to be substituted there). In this type, the two Laguerre orthogonal functions, whose product is to be integrated, *have not the same argument*, but, for example, in our case, have the arguments λ_1/la_0 and $\lambda_1/l'a_0$ where l and l' are the principal quantum numbers of the two levels that we have combined. Let us consider, as typical, the integral

$$J = \int_0^\infty x^p e^{-\frac{\alpha+\beta}{2}x} L_{n+k}^n(\alpha x) L_{n'+k'}^{n'}(\beta x)\, dx. \qquad (116)$$

Now we can proceed in a superficially different way. At first, the former procedure still goes on smoothly; only on the right-hand side of (114) a somewhat more complicated expression appears. In the denominator occurs the power of a quadrinomial instead of that of a binomial, as before. And this makes the matter somewhat confusing, for the right-hand side of (114) becomes five-fold instead of three-fold, and thus the right side of (115) becomes a three-fold instead of a simple sum. I found that the following substitution made things clearer:

$$\frac{\alpha+\beta}{2}x = y. \qquad (117)$$

Hence

$$\begin{cases} \alpha x = \left(1 + \dfrac{\alpha-\beta}{\alpha+\beta}\right) y \\[2mm] \beta x = \left(1 - \dfrac{\alpha-\beta}{\alpha+\beta}\right) y. \end{cases} \qquad (118)$$

After expanding the two polynomials in their Taylor series, which are finite and have similar polynomials as coefficients, we get, using the abbreviations

$$\sigma = \frac{2}{\alpha+\beta}, \qquad \gamma = \frac{\alpha-\beta}{\alpha+\beta}, \qquad (119)$$

the following,

$$J = \sigma^{p+1} \sum_{\lambda=0}^{k} \sum_{\mu=0}^{k'} (-1)^{\mu} \frac{\gamma^{\lambda+\mu}}{\lambda!\,\mu!} \int_0^{\infty} y^{p+\lambda+\mu} L_{n+k}^{n+\lambda}(y) L_{n'+k'}^{n'+\mu}(y) dx.$$

(120)

Thus the calculation of J is reduced to the simpler type of integral (115). In the case of the Balmer lines, the double sum in (120) is comparatively tractable, for one of the two k-values, namely, the one referring to the two-quantum level, never exceeds unity, and thus λ may have two values at most, and, as it turns out, μ four values at most. The circumstance that out of the polynomials referring to the two-quantum level, none but

$$L_0 = 1, \quad L_1 = -x + 1, \quad L_1^1 = -1,$$

appear, permits further simplifications. Nevertheless we must calculate out a number of tables, and it is much to be regretted that the figures given in the tables of the text for the intensities do not allow their general construction to be seen. By good fortune the additive relations between the $\parallel$ and the $\perp$ components hold good, so that we may, with some *probability*, feel ourselves safe from arithmetical blunders at least.

§ 3. Integrals with Legendre Functions

There are three simple integral relations between associated Legendre functions, which are necessary for the calculations in § 5. For the convenience of others, I will state them here, because I was not able to discover them in any of the places I searched. We use the customary definition,

$$P_n^m(\cos\theta) = \sin^m\theta \, \frac{d^m P_n(\cos\theta)}{(d\cos\theta)^m}.$$

(121)

Then the following holds,

$$\int_0^\pi [P_n^m(\cos\theta)]^2 \sin\theta\, d\theta = \frac{2}{2n+1}\frac{(n+m)!}{(n-m)!} \qquad (122)$$

(the normalising relation).

Moreover,

$$\begin{cases} \int_0^\pi P_n^m(\cos\theta)P_{n'}^m(\cos\theta)\cos\theta\sin\theta\, d\theta = 0 \\ \text{for } |n-n'| \neq 1. \end{cases} \qquad (123)$$

On the other hand,

$$\int_0^\pi P_n^m(\cos\theta)P_{n-1}^m(\cos\theta)\cos\theta\sin\theta\, d\theta$$

$$= \frac{n+m}{2n+1}\int_0^\pi [P_{n-1}^m(\cos\theta)]^2\sin\theta\, d\theta \qquad (124)$$

$$= \frac{2(n+m)!}{(4n^2-1)(n-m-1)!}.$$

The last two relations decide the "selection" of the determinantal terms on page 149 of the text. They are, moreover, of fundamental importance for the theory of spectra, for it is obvious that the selection principle for the azimuthal quantum number depends on them (and on two others which have $\sin^2\theta$ in place of $\cos\theta\sin\theta$).

Addition at Proof Correction

Hr. W. Pauli, jun., informs me that he has arrived at the following closed formulae for the total intensity of the lines in the Lyman and Balmer series, through a modification of the method given in section 2 of the Appendix. For the Lyman series these are

$$\nu_{l,1} = R\left(\frac{1}{1^2} - \frac{1}{l^2}\right); \quad J_{l,1} = \frac{2^7 \cdot (l-1)^{2l-1}}{1 \cdot (l+1)^{2l+1}}$$

and for the Balmer series

$$\nu_{l,2} = R\left(\frac{1}{2^2} - \frac{1}{l^2}\right); \ J_{l,2} = \frac{4^3 \cdot (l-2)^{2l-3}}{l \cdot (l+2)^{2l+3}} (3l^2 - 4)(5l^2 - 4).$$

The total emission intensities (square of amplitudes into fourth power of the frequency) are proportional to these expressions, within the series in question. The numbers obtained from the formula for the Balmer series are in complete agreement with those given on p. 143.

Zürich, Physical Institute of the University.
(Received May 10, 1926.)

160

Quantisation as a Problem of Proper Values (Part IV[1])

(*Annalen der Physik* (4), vol. 81, 1926)

ABSTRACT: § 1. Elimination of the energy-parameter from the vibration equation. The real wave equation. Non-conservative systems. § 2. Extension of the perturbation theory to perturbations which explicitly contain the time. Theory of dispersion. § 3. Supplementing § 2. Excited atoms, degenerate systems, continuous spectrum. § 4. Discussion of the resonance case. § 5. Generalisation for an arbitrary perturbation. § 6. Relativistic-magnetic generalisation of the fundamental equations. § 7. On the physical significance of the field scalar.

§ 1. Elimination of the Energy-parameter from the Vibration Equation. The Real Wave Equation. Non-conservative Systems

The wave equation (18) or (18″) of Part II., viz.

$$\nabla^2 \psi - \frac{2(E - V)}{E^2} \frac{\partial^2 \psi}{\partial t^2} = 0 \tag{1}$$

---

[1] Cf. *Ann. d. Phys.* 79, pp. 361, 489; 80, p. 437, 1926 (Parts I., II., III.); further, on the connection with Heisenberg's theory, *ibid.* 79, p. 734 (p. 46).

or

$$\nabla^2\psi + \frac{8\pi^2}{h^2}(E - V)\psi = 0, \qquad (1')$$

which forms the *basis* for the re-establishment of mechanics attempted in this series of papers, suffers from the disadvantage that it expresses the law of variation of the "mechanical field scalar" ψ, neither *uniformly* nor *generally*. Equation (1) contains the energy- or frequency-parameter E, and is valid, as is expressly emphasized in Part II., with a *definite* E-value inserted, for processes which depend on the time exclusively through a *definite* periodic factor:

$$\psi \sim \text{real part of } \left(e^{\pm\frac{2\pi i E t}{h}}\right). \qquad (2)$$

Equation (1) is thus not really any more general than equation (1'), which takes account of the circumstance just mentioned and does not contain the time at all.

Thus, when we designated equation (1) or (1'), on various occasions, as "the wave equation", we were really wrong and would have been more correct if we had called it a "vibration-" or an "amplitude-" equation. However, we found it sufficient, because to *it* is linked the Sturm-Liouville proper value problem – just as in the mathematically strictly analogous problem of the free vibrations of strings and membranes – and not to the *real* wave equation.

As to this, we have always postulated up till now that the potential energy V is a pure function of the coordinates and does *not* depend explicitly on the time. There arises, however, an urgent need for the extension of the theory to *non-conservative* systems, because it is only in that way that we can study the behaviour of a system under the influence of prescribed external forces, e.g. a light wave, or a strange atom flying past. Whenever V contains the time explicitly, it is manifestly *impossible* that equation (1) or (1') should be satisfied by a function ψ, the method of dependence of which on the time is as given by (2). We then find that the amplitude equation is no longer sufficient and that we must search for the real wave equation.

For conservative systems, the latter is easily obtained. (2) is equivalent to

$$\frac{\partial^2 \psi}{\partial t^2} = -\frac{4\pi^2 E^2}{h^2}\psi. \tag{3}$$

We can eliminate E from (1') and (3) by differentiation, and obtain the following equation, which is written in a symbolic manner, easy to understand:

$$\left(\nabla^2 - \frac{8\pi^2}{h^2}V\right)^2 \psi + \frac{16\pi^2}{h^2}\frac{\partial^2 \psi}{\partial t^2} = 0. \tag{4}$$

This equation must be satisfied by every ip which depends on the time as in (2), though *with E arbitrary*, and consequently also by every ψ which can be expanded in a Fourier series with respect to the time (naturally with functions of the coordinates as coefficients). Equation (4) is thus evidently *the uniform and general wave equation for the field scalar ψ*.

It is evidently no longer of the simple type arising for vibrating membranes, but is of the *fourth* order, and of a type similar to that occurring in many problems in the theory of elasticity.[2] However, we need not fear any excessive complication of the theory, or any necessity to revise the previous methods, associated with equation (1'). If V does *not* contain the time, we can, proceeding from (4), apply (2), and then split up the operator as follows:

$$\left(\nabla^2 - \frac{8\pi^2}{h^2}V + \frac{8\pi^2}{h^2}E\right)\left(\nabla^2 - \frac{8\pi^2}{h^2}V - \frac{8\pi^2}{h^2}E\right)\psi = 0. \tag{4'}$$

By way of trial, we can resolve this equation into two "alternative" equations, namely, into equation (1') and into another,

---

[2] E.g., for a vibrating *plate*, $\nabla^2\nabla^2 u + \frac{\partial^2 u}{\partial t^2} = 0$. Cf. Courant-Hilbert, chap. v. § 8, 256.

which only differs from $(1')$ in that its proper value parameter will be *called* minus E, instead of plus E. According to (2) this does not lead to new solutions. The decomposition of $(4')$ is not absolutely cogent, for the theorem that "a product can only vanish when at least *one* factor vanishes" is not valid for operators. This lack of cogency, however, is a feature common to all the methods of solution of partial differential equations. The procedure finds its subsequent justification in the fact that we can prove the *completeness* of the discovered proper functions, as functions of the coordinates. This completeness, coupled with the fact that the imaginary part as well as the real part of (2) satisfies equation (4), allows arbitrary initial conditions to be fulfilled by ψ and $\partial\psi/\partial t$.

Thus we see that the wave equation (4), which contains in itself the law of dispersion, can really stand as the basis of the theory previously developed for conservative systems. The generalisation for the case of a time-varying potential function nevertheless demands caution, because terms with time derivatives of V may then appear, about which no information can be given to us by equation (4), owing to the way we obtained it. In actual fact, if we attempt to apply equation (4) as it stands to non-conservative systems, we meet with complications, which seem to arise from the term in $\partial V/\partial t$. Therefore, in the following discussions, I have taken a somewhat different route, which is much easier for calculations, and which I consider is justified in principle.

We *need* not raise the order of the wave equation to four, in order to get rid of the energy-parameter. The dependence of ψ on the time, which must exist if $(1')$ is to hold, can be expressed by

$$\frac{\partial\psi}{\partial t} = \pm\frac{2\pi i}{h}E\psi \tag{$3'$}$$

as well as by (3). We thus arrive at one of the two equations

$$\nabla^2\psi - \frac{8\pi^2}{h^2}V\psi \mp \frac{4\pi i}{h}\frac{\partial\psi}{\partial t} = 0. \tag{$4''$}$$

We will require the complex wave function ψ to satisfy one of these two equations. Since the conjugate complex function $\bar{\psi}$ will then satisfy the *other* equation, we may take the real part of ψ as the real wave function (if we require it). In the case of a conservative system (4″) is essentially equivalent to (4), as the real operator may be split up into the product of the two conjugate complex operators if V does not contain the time.

§ 2. Extension of the Perturbation Theory to Perturbations containing the Time explicitly. Theory of Dispersion

Our main interest is not in systems for which the time and spatial variations of the potential energy V are of the same order of magnitude, but in systems, conservative in themselves, which are *perturbed* by the addition of small given functions of the time (and of the coordinates) to the potential energy. Let us, therefore, write

$$V = V_0(x) + r(x, t), \qquad (5)$$

where, as often before, x represents the whole of the configuration coordinates. We regard the unperturbed proper value problem ($r = 0$) as *solved*. Then the perturbation problem can be solved by *quadratures*.

However, we will not treat the general problem immediately, but will select the problem of the *dispersion theory* out of the vast number of weighty applications which fall under this heading, on account of its striking importance, which really justifies a separate treatment in any case. Here the perturbing forces originate in an alternating electric field, homogeneous and vibrating synchronously in the domain of the atom; and thus, if we have to do with a linearly polarised monochromatic light of frequency ν, we write

$$r(x, t) = A(x) \cos 2\pi\nu t, \qquad (6)$$

and hence
$$V = V_0(x) + A(x)\cos 2\pi\nu t. \qquad (5')$$

Here $A(x)$ is the negative product of the light-amplitude and the coordinate function which, *according to ordinary mechanics*, signifies the component of the electric moment of the atom in the direction of the electric light-vector (say $-F\sum e_i z_i$, if F is the light-amplitude, $e_i z_i$ the charges and z-coordinates of the particles, and the light is polarised in the z-direction). We borrow the time-*variable* part of the potential function from ordinary mechanics with just as much or as little right as previously, e.g. in the Kepler problem, we borrowed the *constant* part.

Using (5'), equation (4'') becomes

$$\nabla^2\psi - \frac{8\pi^2}{h^2}\left(V_0 + A\cos 2\pi\nu t\right)\psi \mp \frac{4\pi i}{h}\frac{\partial\psi}{\partial t} = 0. \qquad (7)$$

For $A = 0$, these equations are changed by the substitution

$$\psi = u(x)e^{\pm\frac{2\pi iEt}{h}} \qquad (8)$$

(which is now to be taken in the literal sense, and does *not* imply *pars realis*) into the amplitude equation (1') of the unperturbed problem, and we know (cf. § 1) that the totality of the solutions of the unperturbed problem is found in this way. Let
$$E_k \text{ and } u_k(x); \ k = 1, 2, 3, \ldots$$

be the proper values and normalised proper functions of the unperturbed problem, which we regard as *known*, and which we will assume to be *discrete* and *different* from one another (non-degenerate system with no continuous spectrum), so that we may not become involved in secondary questions, requiring special consideration.

Just as in the case of a perturbing potential independent of the time, we will have to seek solutions of the perturbed problem in the neighbourhood of *each* possible solution of the unperturbed problem, and thus in the neighbourhood of an arbitrary linear combination of the u_k's, which has constant coefficients [from (8), the to be combined with the appropriate time factors $e^{\pm\frac{2\pi i E_k t}{h}}$]. The solution of the perturbed problem, lying in the neighbourhood of a *definite* linear combination, will have the following physical meaning. It will be *this solution* which first appears, if, when the light wave arrived, precisely that definite linear combination of free proper vibrations was present (perhaps with trifling changes during the "excitation").

Since, however, the equation of the perturbed problem is also *homogeneous* – let this want of analogy with the "forced vibrations" of acoustics be expressly emphasized – it is evidently sufficient to seek the perturbed solution in the neighbourhood of each *separate*

$$u_k(x)e^{\pm\frac{2\pi i E_k t}{h}},\tag{9}$$

as we may then linearly combine these *ad lib.*, just as for unperturbed solutions.

To solve the first of equations (7) we therefore now put

$$\psi = u_k(x)e^{\frac{2\pi i E_k t}{h}} + w(x,t).\tag{10}$$

[The lower symbol, i.e. the second of equations (7), is henceforth left on one side, as it would not yield anything new.] The additional term $w(x,t)$ can be regarded as small, and its product with the perturbing potential neglected. Bearing this in mind while substituting from (10) in (7), and remembering that $u_k(x)$ and E_k are proper functions and values of the unperturbed problem, we get

$$
\begin{cases}
\nabla^2 w - \dfrac{8\pi^2}{h^2} V_0 w - \dfrac{4\pi i}{h}\dfrac{\partial w}{\partial t} \\[2mm]
= \dfrac{8\pi^2}{h^2} A \cos 2\pi\nu t \cdot u_k\, e^{\frac{2\pi i E_k t}{h}} \\[2mm]
= \dfrac{4\pi^2}{h^2} A u_k \cdot \left(e^{\frac{2\pi i t}{h}(E_k + h\nu)} + e^{\frac{2\pi i t}{h}(E_k - h\nu)} \right).
\end{cases}
\tag{11}
$$

This equation is readily, and really *only*, satisfied by the substitution

$$
w = w_+(x)\, e^{\frac{2\pi i t}{h}(E_k + h\nu)} + w_-(x)\, e^{\frac{2\pi i t}{h}(E_k - h\nu)},
\tag{12}
$$

where the two functions $w_\pm$ respectively obey the two equations

$$
\nabla^2 w_\pm + \frac{8\pi^2}{h^2}(E_k \pm h\nu - V_0) w_\pm = \frac{4\pi^2}{h^2} A u_k.
\tag{13}
$$

This step is essentially *unique*. At first sight, we apparently can add to (12) an arbitrary aggregate of unperturbed proper vibrations. But this aggregate would necessarily be assumed small, of the first order (since this has been assumed for w), and thus does not interest us at present, as it could only produce perturbations of the second order at most.

In equations (13) we have at last those *non-homogeneous* equations we might have expected to encounter – in spite of the lack of analogy with real forced vibrations, as emphasized above. This lack of analogy is extraordinarily important and manifests itself in equations (13) in the two following particulars. *Firstly*, as the "second member" ("exciting force"), the perturbation function $A(x)$ does not appear *alone* but *multiplied* by the amplitude of the free vibration already present. This is indispensable if the physical facts are to be properly taken into account, for the reaction of an atom to an incident light wave depends almost entirely on the *state* of the atom at that time, whereas the forced vibrations of a membrane, plate, etc., are known to be quite independent of the proper vibrations which may be superimposed on them, and thus would yield an obviously wrong representation of our case. *Secondly*, in place of the proper value on the left-hand side of (13), i.e. as "exciting frequency", we do not find the frequency ν of the perturbing force *alone*, but rather in one case added to, and in the other subtracted from, that of the free vibration already present. This is equally indispensable. Otherwise the proper frequencies *themselves*, which correspond to the *term-frequencies*, would function as *resonance-points*, and not the *differences* of the proper frequencies, as is demanded, and is really given by equation (13). Moreover, we see with satisfaction that the latter gives *only* the differences between a proper frequency *which is actually excited* and all the others, and *not* the differences between pairs of proper frequencies, of which *no member* is excited.

In order to investigate this more closely, let us complete the solution. By well-known methods[3] we find, as *simple* solutions of (13),

$$w_{\pm}(x) = \frac{1}{2} \sum_{n=1}^{\infty} \frac{a'_{kn} u_n(x)}{E_k - E_n \pm h\nu}, \tag{14}$$

where

$$a'_{kn} = \int A(x) u_k(x) u_n(x) \rho(x) dx. \tag{15}$$

$\rho(x)$ is the "density function", i.e. that function of the position-coordinates with which equation (1′) must be multiplied to make it self-adjoint. The $u_n(x)$'s are assumed to be normalised. It is further postulated that $h\nu$ *does not agree exactly with any of the differences* $E_k - E_n$ *of the proper values*. This "resonance case" will be dealt with later (cf. § 4).

---

[3]Cf. Part III. §§ I and 2, text beside equations (8) and (24).

If we now form from (14), using (12) and (10), the entire perturbed vibration, we get

$$
\begin{cases}
\psi = u_k(x)e^{\frac{2\pi i E_k t}{h}} \\
+\dfrac{1}{2}\displaystyle\sum_{n=1}^{\infty} a'_{kn}u_n(x)\left(\dfrac{e^{\frac{2\pi i t}{h}(E_k+h\nu)}}{E_k - E_n + h\nu} + \dfrac{e^{\frac{2\pi i t}{h}(E_k-h\nu)}}{E_k - E_n - h\nu} \right).
\end{cases}
\tag{16}
$$

Thus in the perturbed case, along with each *free* vibration $u_k(x)$ occur in small amplitude all those vibrations $u_n(x)$, for which $a'_{kn} \neq 0$. The latter are exactly those, which, if they exist as free vibrations along with u_k, give rise to a radiation, which is (wholly or partially) polarised in the direction of polarisation of the incident wave. For apart from a factor, a'_{kn} is just the component amplitude, in this direction of polarisation, of the atom's *electric moment*, which is oscillating with frequency $(E_k - E_n)/h$, *according to wave mechanics*, and which appears when u_k and u_n exist together.[4] The simultaneous oscillation, however, takes place with neither the proper frequency E_n/h, peculiar to these vibrations, nor the frequency ν of the light wave, but rather with the sum and difference of ν and E_k/h (i.e. the frequency of the *one* existing *free* vibration).

The real or the imaginary part of (16) can be considered as the *real* solution. In the following, however, we will operate with the complex solution itself.

To see the significance that our result has in the theory of dispersion, we must examine the radiation arising from the simultaneous existence of the excited forced vibrations and the free vibration, already present. For this purpose, we form, following the method we[5] have always adopted above –

---

[4] Cf. what follows, and § 7.

[5] Cf. end of paper on Quantum Mechanics of Heisenberg, etc., and also the Calculation of Intensities in the Stark Effect in Part III. At the first quoted place, the real part of $\psi\bar{\psi}$ was proposed instead of $\psi\bar{\psi}$. This was a mistake, which was corrected in Part III.

a criticism follows in § 7 – the product of the complex wave function (16) and its conjugate, i.e. the norm of the complex wave function ψ. We notice that the perturbing terms are small, so that squares and products may be neglected. After a simple reduction[6] we obtain

$$\psi\bar{\psi} = u_k(x)^2 + 2\cos 2\pi\nu t \sum_{n=1}^{\infty} \frac{(E_k - E_n)a'_{kn}u_k(x)u_n(x)}{(E_k - E_n)^2 - h^2\nu^2}. \quad (17)$$

According to the *heuristic hypothesis* on the electrodynamical significance of the field scalar ψ, the present quantity – apart from a multiplicative constant – represents the electrical density as a function of the space coordinates and the time, *if x stands for only three space coordinates*, i.e. if we are dealing with the problem of *one* electron. We remember that the same hypothesis led us to correct selection and polarisation rules and to a very satisfactory representation of intensity relationships in our discussion of the hydrogen Stark effect. By a natural generalisation of this hypothesis – of which more in § 7 – we regard the following as representing in the general case the density of the electricity, which is "associated" with *one* of the particles of classical mechanics, or which "originates in it", or which "corresponds to it in wave mechanics": the *integral* of $\psi\bar{\psi}$ taken over all those coordinates of the system, which in classical mechanics fix the position of *the rest of the* particles, multiplied by a certain constant, the classical "charge" of the first particle. The resultant density of charge at any point of space is then represented by the *sum* of such integrals taken over all the particles.

Thus in order to find any space component whatever of the total wave-mechanical dipole moment as a function of the

---

[6]We assume as previously, for the sake of simplicity, the proper functions $u_n(x)$ to be *real*, but notice that it may sometimes be much more convenient or even imperative to work with complex aggregates of the real proper functions, e.g. in the proper functions of the Kepler problem to work with $e^{\pm m\phi i}$ instead of $\begin{smallmatrix}\cos\\\sin\end{smallmatrix} m\phi$.

time, we must, on this hypothesis, multiply expression (17) by that function of the coordinates which gives that particular dipole component in *classical mechanics* as a function of the configuration of the point system, e.g. by

$$M_y = \sum e_i y_i, \tag{18}$$

if we are dealing with the dipole moment in the y-direction. Then we have to integrate over *all* the configuration coordinates.

Let us work this out, using the abbreviation

$$b_{kn} = \int M_y(x)u_k(x)u_n(x)\rho(x)dx. \tag{19}$$

Let us elucidate further the definition (15) of the a'_{kn}'s by recalling that if the incident electric light-vector is given by

$$\mathfrak{E}_z = F\cos 2\pi\nu t, \tag{20}$$

then

$$\begin{cases} A(x) = -F \cdot M_z(x), \\ \text{where } M_z(x) = \sum e_i z_i. \end{cases} \tag{21}$$

If we put, in analogy with (19),

$$a_{kn} = \int M_z(x)u_k(x)u_n(x)\rho(x)dx, \tag{22}$$

then $a'_{kn} = -Fa_{kn}$, and by carrying out the proposed integration we find,

$$\int M_y\psi\bar{\psi}\rho dx = a_{kk} + 2F\cos 2\pi\nu t \sum_{n=1}^{\infty} \frac{(E_n - E_k)a_{kn}b_{kn}}{(E_k - E_n)^2 - h^2\nu^2} \tag{23}$$

for *the resulting electric moment, to which the secondary radiation, caused by the incident wave* (20), *is to be attributed.*

The radiation depends of course only upon the second (time-variable) part, while the first part represents the time-constant dipole moment, which is possibly connected with the originally existing free vibration. This variable part seems fairly promising and may meet all the demands we are accustomed to make on a "dispersion formula". Above all, let us note the appearance of those so-called "negative" terms, which – in the usual phraseology – correspond to the probability of transition to a lower level ($E_n < E_k$), and to which Kramers[7] was the first to direct attention, from a correspondence standpoint. Generally, our formula – despite very different ways of thought and expression – may be characterised as really identical in form with Kramer's formula for secondary radiation. The important connection between a_{kn}, b_{kn}, the coefficients of the secondary and of the spontaneous radiation, is brought out, and indeed the secondary radiation is also described accurately with respect to its condition of polarisation.[8]

I would like to believe that the absolute value of the scattered radiation or of the induced dipole moment is also given correctly by formula (23), although it is obviously within the bounds of possibility that an error in the numerical factor may have occurred in applying the heuristic hypothesis introduced above. At any rate the physical dimensions are right, for from (18), (19), (21), and (22) a_{kn} and b_{kn} are electric moments, since the squared integrals of the proper functions were normalised to unity. If ν is far removed from the emis-

---

[7]H. A. Kramers, *Nature*, May 10, 1924; *ibid.* August 30, 1924; Kramers and W. Heisenberg. *Ztschr. f. Phys.* 31, p. 681, 1925. The description given in the latter paper of the polarisation of the scattered light (equation 27) from correspondence principles, is almost identical *formally* with ours.

[8]It is hardly necessary to say that the two directions which, for simplicity, we have designated as "z-direction" and "y-direction" do not require to be exactly perpendicular to one another. The one is the direction of polarisation of the incident wave; the other is that polarisation component of the secondary wave, in which we are specially interested.

sion frequency in question, the ratio of the induced to the spontaneous dipole moment is of the same order of magnitude as the ratio of the additional potential energy Fa_{kn} to the "energy step" $E_k - E_n$.

§ 3. Supplements to § 2. Excited Atoms, Degenerate Systems, Continuous Spectrum

For the sake of clearness, we have made some special assumptions, and put many questions aside, in the preceding paragraph. These have now to be discussed by way of supplement.

First: what happens when the light wave meets the atom, when the latter is in a state in which not merely *one* free vibration, u_k, is excited as hitherto assumed, but several, say two, u_k and u_l? As remarked above, we have in the perturbed case simply to combine additively the two perturbed solutions (16) corresponding to the suffix k and the suffix l, after we have provided them with constant (possibly complex) coefficients, which correspond to the *strength* presumed for the free vibrations, and to the phase relationship of their stimulation. Without actually performing the calculation, we see that in the expression for $\psi\bar{\psi}$ and also in the expression (23) for the resulting electric moment, there then occurs *not merely* the corresponding linear aggregate of the terms previously obtained, i.e, of the expressions (17) or (23) written with k, and then with l. We have in addition "combination terms", namely, considering *first* the greatest order of magnitude, a term in

$$u_k(x)u_l(x)e^{\frac{2\pi i}{h}(E_k-E_i)t}, \tag{24}$$

which gives again the *spontaneous* radiation, bound up with the co-existence of the two *free* vibrations; and *secondly* perturbing terms of the first order, which are proportional to the perturbing field amplitude, and which correspond to the interaction of the forced vibrations belonging to u_k with the free vibration u_l – and of the forced vibrations belonging to u_l with u_k. The *frequency* of these new terms appearing in (17) or (23) is *not* ν but

$$\left| \nu \pm \frac{(E_k - E_l)}{h} \right|, \tag{25}$$

as can easily be seen, still without carrying out the calculation. (New "resonance denominators", however, do *not* occur in these terms.) Thus we have to do here with a secondary radiation, whose frequency neither coincides with the exciting light-frequency nor with a spontaneous frequency of the system, but is a combination frequency of both.

The existence of this remarkable kind of secondary radiation was first postulated by Kramers and Heisenberg (*loc. cit.*), from correspondence considerations, and then by Born, Heisenberg, and Jordan from consideration of Heisenberg's quantum mechanics.[9] As far as I know, it has not yet been demonstrated experimentally. The *present* theory also shows distinctly that the occurrence of this scattered radiation is dependent on special conditions, which demand researches expressly arranged for the purpose. Firstly, *two* proper vibrations u_k and u_l must be *strongly* excited, so that all experiments made on atoms in their normal state – as happens in the vast majority of cases – are to be rejected. Secondly, at least *one* third state of proper vibration must *exist* (i.e. must be possible – it need not be *excited*), which leads to powerful spontaneous emission, when combined with u_k as well as with u_l. For the extraordinary scattered radiation, which is to be discovered, is proportional to the product of the spontaneous emission coefficients in question ($a_{kn}b_{ln}$ and $a_{ln}b_{kn}$). The combination (u_k, u_l) need not, in itself, cause a strong emission. It would not matter if – to use the language

[9]Bom, Heisenberg, and Jordan, *Ztschr. f. Phys.* 35, p. 572, 1926.

of the older theory – this was a "forbidden transition". Yet in practice we must also demand that the line (u_k, u_l) should actually be emitted strongly during the experiment, for this is the only means of assuring ourselves that *both* proper vibrations are strongly excited in the same individual atoms and in a sufficiently great number of them. If we reflect now that in the powerful term-series mostly examined, i.e . in the ordinary s-, p-, d-, f-series, the relations are generally such that two terms, which combine strongly with a third, do not do so with one another, then a special choice of the object and conditions of the research seems really necessary, if we are to expect the desired scattered radiation with any certainty, especially as its frequency is not that of the exciting light and *thus* it does not produce dispersion or rotation of the plane of polarisation, but can only be observed as light scattered on all sides.

As far as I see, the above - mentioned dispersion theory of Heisenberg, Born, and Jordan does *not* allow of such reflections as we have just made, in spite of its great formal similarity to the present one. For it only considers *one* way in which the atom reacts to incident radiation. It conceives the atom as a timeless entity, and up till now is not able to express in its language the undoubted fact that the atom can be in *different* states at different times, and thus, as has been proved, reacts in different ways to incident radiation.[10]

Let us turn now to another question. In § 2 the collective proper values were postulated to be *discrete* and *different* from one another. We now drop the second hypothesis and ask: what is altered when *multiple* proper values occur, i.e. when *degeneracy* is present? Perhaps we expect that complications then arise, similar to those we met in the case of a time-constant perturbation (Part III. § 2), i.e. that a system

---

[10]Cf. especially the concluding words of Heisenberg's latest exposition of his theory, *Math. Ann.* 95, p. 683, 1926, in connection with this difficulty of comprehending *the course of an event in time*.

of proper functions of the unperturbed atom, suited to the particular perturbation, must be defined by the solution of a "secular equation", and applied to carry out the perturbation calculation. This is indeed so in the case of an *arbitrary* perturbation, represented by $r(x,t)$ as in equation (5), but *not* so in the case of a perturbation by a light wave (equation (6)) – at any rate, for our usual first approximation, and as long as we suppose that the light frequency ν does not coincide with any of the spontaneous emission frequencies considered. Then the parameter value in the double equation (13), for the amplitudes of the perturbed vibrations, is *not* a proper value, and the pair of equations has always the unambiguous pair of solutions (14), in which no vanishing denominators occur even when E_k is a multiple value. Thus the terms in the sum for which $E_n = E_k$ are *not, as might be thought*, to be omitted, any more than the term for $n = k$ itself. It is worth noticing that through these terms – if one of them occurs really, i.e. with non-vanishing a_{kn} – the frequency $\nu = 0$ also appears among the resonance frequencies. These terms do *not*, of course, contribute to the "ordinary" scattered radiation, as we see from (23), since $E_k - E_n = 0$.

The simplification, that we do not require to consider specially any possible degeneracy present, at least in a first approximation, is always available[11] when the time-averaged value of the perturbation function vanishes, or what is the same thing, when the latter's Fourier expansion in terms of the time contains no constant, i.e. time-independent, term. This is the case for a light wave.

While our *first* postulation about the proper values – that they should be *simple* – has thus shown itself to be really a superfluous precaution, a dropping of the *second* – that they should be absolutely discrete – while leading to no alterations *in principle*, brings about, however, very considerable alter-

---

[11]Further discussed in § 5.

178

ations in the external appearance of the calculation, inasmuch as *integrals* taken over the continuous spectrum of equation (1') are to be added to the discrete sums in (14), (16), (17), and (23). The theory of such representations by integrals has been developed by H. Weyl,[12] and though only for ordinary differential equations, the extension to partials is permissible. In all brevity, the state of the case is this.[13] If the homogeneous equation belonging to the non-homogeneous equations (13), i.e. the vibration equation (1') of the unperturbed system, possesses in addition to a point-spectrum a continuous one, which stretches, say, from $E = a$ to $E = b$, then an arbitrary function $f(x)$ naturally cannot be developed thus,

$$f(x) = \sum_{n=1}^{\infty} \phi_n \cdot u_n(x),$$

$$\text{where} \quad \phi_n = \int f(x) u_n(x) \rho(x) dx \tag{26}$$

in terms of the normalised discrete proper functions $u_n(x)$ alone, but there must be added an integral expansion in terms of the proper solutions $u(x, E)$, which belong to the proper values $a \leq E \leq b$, and so we have

$$f(x) = \sum_{n=1}^{\infty} \phi_n \cdot u_n(x) + \int_a^b u(x, E) \phi(E) dE, \tag{27}$$

where to emphasize the analogy we have intentionally chosen the same letter for the "coefficient function" $\phi(E)$ as for the discrete coefficients ϕ_n. If now we have *normalised*, once

---

[12]H. Weyl, *Math. Ann.* 68, p. 220, 1910; *Gött. Nachr.* 1910. Cf. also E. Hilb, *Sitz.-Ber. d. Physik. Mediz. Soc. Erlangen*, 43, p. 68, 1911; *Math. Ann.* 71, p. 76, 1911. I have to thank Herr Weyl not only for these references but also for very valuable oral instruction in these not very simple matters.

[13]I have to thank Herr Fues for this exposition.

for all, the proper solution $u(x, E)$ by associating with it a suitable function of E, in such a way that

$$\int dx \rho(x) \int_{E'}^{E'+\Delta} u(x, E)u(x, E')dE' = 1 \text{ or } 0 \qquad (28)$$

according to whether E *belongs* to the interval $E', E' + \Delta$ or not, then in (27) under the integral sign we substitute from

$$\phi(E) = \lim_{\Delta=0} \frac{1}{\Delta} \int \rho(\xi)f(\xi) \cdot \int_{E}^{E+\Delta} u(\xi, E')dE' \cdot d\xi, \qquad (29)$$

wherein the *first* integral sign refers as always to the domain of the group of variables x.[14] Assuming (28) to be fulfilled and expansion (27) to exist – which statements are proved by Weyl for ordinary differential equations – the definition of the "coefficient functions" from (29) is almost as obvious as the well-known definition of the Fourier coefficients.

The most important and difficult task in any concrete case is the carrying out of the normalisation of $u(x, E)$, i.e. the finding of that function of E by which we have to multiply the (as yet not normalised) proper solution of the continuous spectrum, in order that condition (28) may be satisfied. The above-quoted works of Herr Weyl contain very valuable guidance for this practical task, and also some worked-out examples. An example from atomic dynamics on the intensities of band spectra is worked out by Herr Fues in a paper appearing in the present issue of *Annalen der Physik*.

Let us apply this to our problem, i.e. to the solution of the pair of equations (13) for the amplitudes $w_\pm$ of the perturbed vibrations, where we postulate as usual that the *one* excited *free* vibration, u_k, belongs to the discrete point-spectrum. We

---

[14]As Herr E. Fues informs me, we can very often omit the limiting process in practice and write $u(\xi, E)$ for the inner integral, viz. always, when $\int \rho(\xi)f(\xi)u(\xi, E)d\xi$ exists.

develop the right-hand side of (13) according to the scheme (27) thus,

$$\frac{4\pi^2}{h^2} A(x)u_k(x) = \frac{4\pi^2}{h^2} \sum_{n=1}^{\infty} a'_{kn} u_n(x)$$

$$+ \frac{4\pi^2}{h^2} \int_a^b u(x, E)\alpha'_k(E)dE$$

(30)

in which a'_{kn} is given by (15), and $\alpha'_k(E)$ from (29) by

$$\alpha'_k(E) = \lim_{\Delta=0} \frac{1}{\Delta} \int \rho(\xi)A(\xi)u_k(\xi) \cdot \int_E^{E+\Delta} u(\xi, E')dE' \cdot d\xi.$$

(15')

If we imagine expansion (30) put into (13), and then expand also the desired solution $w_\pm(x)$ similarly in terms of the proper solutions $u_n(x)$ and $u(x, E)$, and notice that for the last-named functions the left side of (13) takes the value

$$\frac{8\pi^2}{h^2} (E_k \pm h\nu - E_n) u_n(x)$$

or

$$\frac{8\pi^2}{h^2} (E_k \pm h\nu - E) u(x, E),$$

then by "comparison of coefficients" we obtain as the generalisation of (14)

$$w_\pm(x) = \frac{1}{2} \sum_{n=1}^{\infty} \frac{a'_{kn} u_n(x)}{E_k - E_n \pm h\nu} + \frac{1}{2} \int_a^b \frac{\alpha'_k(E)u(x, E)}{E_k - E \pm h\nu} dE.$$

(14')

The further procedure is completely analogous to that of § 2. Finally, we get as *additional* term for (23)

$$+ 2\cos 2\pi\nu t \int d\xi \rho(\xi) M_y(\xi) u_k(\xi)$$

$$\times \int_a^b \frac{(E_k - E)\alpha'_k(E)u(\xi, E)}{(E_k - E)^2 - h^2\nu^2} dE.$$

(23')

Here, perhaps, we may not always change the order of integration without further examination, because the integral with respect to ξ may possibly not converge. However, we can – as an intuitive makeshift for a strict passage to the limit, which maybe dispensed with here – decompose the integral $\int_a^b$ into many small parts, each having a range Δ, which is sufficiently small to allow us to regard all the functions of E in question as constant in each part, with the exception of $u(x, E)$, for we know from the general theory that its integral cannot be obtained through such a fixed partition, which is independent of ξ. We can then take the remaining functions out of the partial integrals, and as *additional term for the dipole moment (23) of the secondary radiation*, obtain finally exactly the following,

$$2F \cos 2\pi \nu t \int_a^b \frac{(E_k - E)\alpha_k(E)\beta_k(E)}{(E_k - E)^2 - h^2\nu^2} dE, \qquad (23'')$$

where

$$\alpha_k(E) = \lim_{\Delta=0} \frac{1}{\Delta} \int \rho(\xi) M_z(\xi) u_k(\xi) \cdot \int_E^{E+\Delta} u(\xi, E')dE' \cdot d\xi,$$
$$(22')$$
$$\beta_k(E) = \lim_{\Delta=0} \frac{1}{\Delta} \int \rho(\xi) M_y(\xi) u_k(\xi) \cdot \int_E^{E+\Delta} u(\xi, E')dE' \cdot d\xi$$
$$(19')$$

(please note the complete analogy with the formulae with the same numbers but without the dashes in § 2).

The preceding sketch of the calculation is of course only a general outline, given merely to show that the much-discussed influence of the continuous spectrum on dispersion, which experiment[15] appears to indicate as existing, is required by the present theory exactly in the form expected, and to outline

---

[15]K. F. Herzfeld and K. L. Wolf, *Ann. d. Phys.* 76, p. 71, 667, 1925; R. Kollmann and H. Mark, *Die Nw.* 14, p. 648, 1926.

the way in which the calculation of the problem is to be tackled.

§ 4. Discussion of the Resonance Case

Up till now we have always assumed that the frequency ν of the light wave does not agree with any of the emission frequencies that have to be considered. We now assume that, say,

$$h\nu = E_n - E_k > 0, \tag{31}$$

and we revert, moreover, to the limiting conditions of § 2 for the sake of simplicity (simple, discrete proper values, one single free vibration u_k excited). In the pair of equations (13), the proper value parameter then takes the values

$$E_k \pm E_n \mp E_k = \begin{cases} E_n \\ 2E_k - E_n, \end{cases} \tag{32}$$

i.e. for the upper sign there appears a *proper value*, namely, E_n. The two cases are possible. Firstly, the right side of equation (13) multiplied by $\rho(x)$, may be *orthogonal* bo the proper function $u_n(x)$ corresponding to E_n i.e. we have

$$\int A(x)u_k(x)u_n(x)\rho(x)dx = a'_{kn} = 0, \tag{33}$$

which means, physically, that if u_k and u_n exist together as free vibrations they will give rise to no spontaneous emission or to one which is polarised perpendicularly to the direction of polarisation of the incident light. In this case the critical equation (13) also again possesses a solution, which now, as before, is given by (14), in which the catastrophic term vanishes. This means physically – in the old phraseology – that a "forbidden transition" cannot be stimulated through resonance, or that a "transition", even if not forbidden, cannot be caused by light which is vibrating perpendicularly to the direction of polarisation of that light which would be emitted by the "spontaneous transition".

Otherwise, secondly, (33) is *not* fulfilled. Then the critical equation possesses *no* solution. Statement (10), which assumes a vibration which differs very *little* – by quantities of the order of the light amplitude F – from the originally existing free vibration, and is the *most general* possible *under this assumption, thus does not then lead to the goal.* No solution, therefore, exists which only differs by quantities of the order of F from the original free vibration. The incident light has thus *a varying influence* on the state of the system, *which bears no relation to the magnitude of the light amplitude.* What influence? We can judge this, still without further calculation, if we start out from the case where the resonance condition (31) is not exactly but only approximately fulfilled. Then we see from (16) that $u_n(x)$ is excited in unusually strong forced vibrations, on account of the small denominator, and that – not less important – the frequency of these forced vibrations approaches the natural proper frequency E_n/h of the proper vibration u_n. (All this is, indeed, very *similar* to, yet in a way of its own *different* from, the resonance phenomena encountered elsewhere; otherwise I would not discuss it so minutely.)

In a gradual approach to the critical frequency, the proper vibration u_n, formerly not excited, whose possible existence is responsible for the crisis, is stimulated to a stronger and stronger degree, and with a frequency more and more closely approaching its own proper frequency. In contradistinction to ordinary resonance phenomena there comes a point, and that even before the critical frequency is reached, where our solution does not represent the circumstances correctly any longer, even under the assumption that our obviously "undamped" wave postulation is strictly correct. For we have in fact regarded the forced vibration w as small compared with the existing free vibration and neglected a squared term (in equation (11)).

I believe that the present discussion has already shown, with sufficient clearness, that in the resonance case the theory will actually give the result it ought to give, in order to agree with Wood's resonance phenomenon: an increase of the proper vibration u_n, which causes the crisis, to a finite magnitude comparable with that of the originally existing u_k, from which, of course, "spontaneous emission" of the spectral line (u_k, u_n) results. I do not wish, however, to attempt to work out the calculation of the resonance case fully here, because the result would be of little value, so long as the *reaction* of the emitted radiation on the emitting system is not taken into account. Such a reaction must exist, not only because there is no ground at all for differentiating on principle between the light wave which is incident from outside, and that which is emitted by the system itself, but also because otherwise, if several proper vibrations were simultaneously excited in a system left to itself, the spontaneous emission would continue indefinitely. This required back-coupling must act so that in this case, along with the light emission, the higher proper vibrations gradually die down, and, finally, the fundamental vibration, corresponding to the normal state of the system, alone remains. The back-coupling is evidently exactly analogous to the reaction of radiation $\left(\frac{2e^2}{3mc^3} \ddot{v} \right)$ in the classical electron theory. This analogy also allays the increasing apprehension caused by the previous neglect of this back-coupling. The influence of the relevant term (probably no longer linear) in the wave equation will generally be small, just as in the electron the back pressure of radiation is generally very small compared with the force of inertia and the external field strength. In the resonance case, however — just as in the electron theory – the coupling with the proper light wave will be of the same order as that with the incident wave, and must be taken into account, if the "equilibrium" between the different proper vibrations, which sets in for the given irradiation, is to be correctly computed.

Let it be expressly remarked, however, that the back-coupling term is *not necessary for averting a resonance catas-trophe*! Such can never occur in any circumstances, because according to the theorem of the *persistence of normalisation*, proved below in § 7, the configuration space integral of $\psi\bar{\psi}$ always remains normalised to the same value, even under the influence of arbitrary external forces – and indeed quite au-tomatically, as a consequence of the wave equation $(4'')$. The amplitudes of the ψ-vibrations, therefore, cannot grow indef-initely; they have, "on the average", always the same value. If *one* proper vibration waxes, then another must, therefore, wane.

§ 5. Generalisation for an Arbitrary Perturbation

If an *arbitrary* perturbation is in question as was assumed in equation (5) at the beginning of § 2, then we shall expand the perturbation energy $r(x,t)$ as a Fourier series or Fourier integral in terms of the time. The terms of this expansion have, then, the form (6) of the perturbation potential of a light wave. We see immediately that on the right-hand side of equation (11) we then simply get two *series* (or, possibly, integrals) of imaginary powers of e, instead of merely two terms. If none of the exciting frequencies coincide with a critical frequency, we get the solution in exactly the same way as described in § 2, but, naturally, as Fourier series (or possibly Fourier integrals) of the time. It serves no purpose to write down the formal expansions here, and a more exact working out of separate problems lies outside the scope of the present paper. Yet an important point, already touched upon in § 3, must be mentioned.

Among the critical frequencies of equation (13), the fre-quency $\nu = 0$, from $E_k - E_k = 0$, also generally figures. For in this case also one proper value, namely, E_k appears on the left side as proper value parameter. Thus, if the frequency 0, i.e. a term independent of the time, occurs in the Fourier expansion of the perturbation function $r(x,t)$, we cannot reach our goal by exactly the earlier method. We easily see, however, how it must be modified, for the case of a time-constant perturbation is known from previous work (cf. Part III.). We have then to consider, at the same time, a small alteration and possibly a splitting up of the proper value or values of the excited free vibrations, i.e. in the indices of the powers of e in the first term on the right hand of equation (10) we have to replace E_k by E_k plus a small constant, the perturbation of the proper value. Exactly as described in Part III., § 1 and § 2, this perturbation is defined by the postulation that the right side of the critical Fourier component of our equation (13) is to be orthogonal to (or possibly to *all* the proper functions belonging to E_k).

The number of special problems, which fall under the question formulated in the present paragraph, is extraordinarily great. By superposing the perturbations due to a constant electric or magnetic field and a light wave, we obtain magnetic and electric double refraction, and magnetic rotation of the plane of polarisation. Resonance radiation in a magnetic field also comes under this heading, but for this purpose we must first obtain an exact solution for the resonance case discussed in § 4. Further, we can treat the action of an α-particle or electron flying past the atom[16] in this way, if the encounter is not too close for the perturbation of each of the two systems to be calculable from the undisturbed motion of the other. All these questions are mere matters of calculation as soon as the proper values and functions of the unperturbed systems

---

[16] A very interesting and successful attempt to compare the action of flying charged particles with the action of light waves, through a Fourier decomposition of their field, is to be found in a paper by E. Fermi, *Ztschr. f. Phys.* 29, p. 315, 1924.

are known. It is, therefore, to be hoped that we will succeed in defining these functions, at least approximately, for heavier atoms also, in analogy with the approximate definition of the Bohr electronic orbits which belong to different types of terms.

§ 6. Relativistic-magnetic Generalisation of the Fundamental Equations

As an appendix to the physical problems just mentioned, in which the magnetic field, which has hitherto been completely ignored in this series of papers, plays an important part, I would like to give, briefly, the probable relativistic-magnetic generalisation of the basic equations (4''), although I can only do this meantime for the one electron problem, and only with the greatest possible reserve – the latter for two reasons. *Firstly*, the generalisation is provisionally based on a purely formal analogy. *Secondly*, as was mentioned in Part I., though it does formally lead in the Kepler problem to Sommerfeld's fine-structure formula with, in fact, the "half-integral" azimuthal and radial quantum, which is generally regarded as correct today, nevertheless there is *still lacking* the *supplement*, which is necessary to secure numerically correct diagrams of the splitting up of the hydrogen lines, and which is given in Bohr's theory by Goudsmit and Uhlenbeck's electronic spin.

The Hamilton-Jacobi partial differential equation for the Lorentzian electron can readily be written:

$$\left\{ \begin{aligned} &\left(\frac{1}{c}\frac{\partial W}{\partial t} + \frac{e}{c}V\right)^2 - \left(\frac{1}{c}\frac{\partial W}{\partial x} - \frac{e}{c}\mathfrak{U}_x\right)^2 - \left(\frac{1}{c}\frac{\partial W}{\partial y} - \frac{e}{c}\mathfrak{U}_y\right)^2 \\ &- \left(\frac{1}{c}\frac{\partial W}{\partial z} - \frac{e}{c}\mathfrak{U}_z\right)^2 - m^2c^2 = 0. \end{aligned} \right.$$

(34)

Here e, m, c are the charge and mass of the electron, and the velocity of light; V, $\mathfrak{U}$ are the electro-magnetic potentials of the external electro-magnetic field at the position of the electron, and W is the action function.

From the classical (relativistic) equation (34) I am now attempting to derive the *wave equation* for the electron, by the following *purely formal* procedure, which, we can verify easily, will lead to equations (4″), if it is applied to the Hamiltonian equation of a particle moving in an arbitrary field of force in ordinary (non-relativistic) mechanics. *After* the squaring, in equation (34), I replace the *quantities*

$$
\begin{cases}
\dfrac{\partial W}{\partial t}, \quad \dfrac{\partial W}{\partial x}, \quad \dfrac{\partial W}{\partial y}, \quad \dfrac{\partial W}{\partial z}, \\[2mm]
\text{by the respective } operators \\[2mm]
\pm\dfrac{h}{2\pi i}\dfrac{\partial}{\partial t}, \quad \pm\dfrac{h}{2\pi i}\dfrac{\partial}{\partial x}, \quad \pm\dfrac{h}{2\pi i}\dfrac{\partial}{\partial y}, \quad \pm\dfrac{h}{2\pi i}\dfrac{\partial}{\partial z}.
\end{cases}
\tag{35}
$$

The double linear operator, so obtained, is applied to a wave function ψ and the result put equal to zero, thus:

$$
\nabla^2\psi - \frac{1}{c^2}\frac{\partial^2\psi}{\partial t^2} \mp \frac{4\pi i e}{hc}\left(\frac{V}{c}\frac{\partial\psi}{\partial t} + \mathfrak{U}\,\mathrm{grad}\,\psi\right)
$$
$$
+ \frac{4\pi^2 e^2}{h^2 c^2}\left(V^2 - \mathfrak{U}^2 - \frac{m^2 c^4}{e^2}\right)\psi = 0.
\tag{36}
$$

(The symbols ∇^2 and grad have here their elementary three-dimensional Euclidean meaning.) The pair of equations (36) would be the possible relativistic-magnetic generalisation of (4″) for the case of a single electron, and should likewise be understood to mean that the complex wave function has to satisfy either the one or the other equation.

From (36) the fine structure formula of Sommerfeld for the hydrogen atom may be obtained by exactly the same method as is described in Part I., and also we may derive (neglecting the term in $\mathfrak{U}^2$) the normal Zeeman effect as well as the well-known selection and polarisation rules and intensity formulae. They follow from the integral relations between Legendre functions introduced at the end of Part III.

For the reasons given in the first section of this paragraph, I withhold the detailed reproduction of these calculations meantime, and also in the following final paragraph refer to the "classical", and not to the still incomplete relativistic-magnetic version of the theory.

§ 7. On the Physical Significance of the Field Scalar

The heuristic hypothesis of the electro-dynamical meaning of the field scalar ψ, previously employed in the *one*-electron problem, was extended off-hand to an arbitrary system of charged particles in § 2, and there a more exhaustive description of the procedure was promised. We had calculated the density of electricity at an arbitrary point in space as follows. We selected *one* particle, kept the trio of coordinates that describes *its* position in ordinary mechanics fixed; integrated $\psi\bar{\psi}$ over all the rest of the coordinates of the system and multiplied the result by a certain constant, the "charge" of the selected particle; we did a similar thing for each particle (trio of coordinates), in each case giving the selected particle the same position, namely, the position of that point of *space* at which we desired to know the electric density. The latter is equal to the algebraic sum of the partial results.

This rule is now equivalent to the following conception, which allows the true meaning of ψ to stand out more clearly, $\psi\bar{\psi}$ is a kind of *weight-function* in the system's configuration space. The *wave-mechanical* configuration of the system is a *superposition* of many, strictly speaking of *all*, point-mechanical configurations kinematically possible. Thus, each point-mechanical configuration contributes to the true wave-mechanical configuration with a certain *weight*, which is given precisely by $\psi\bar{\psi}$. If we like paradoxes, we may say that the system exists, as it were, simultaneously in all the positions kinematically imaginable, but not "equally strongly" in all. In macroscopic motions, the weight-function is practically concentrated in a small region of positions, which are practically indistinguishable. The centre of gravity of this region in configuration space travels over distances which are macroscopically perceptible. In problems of microscopic motions, we are in any case interested *also*, and in certain cases even *mainly*, in the varying *distribution* over the region.

This new interpretation may shock us at first glance, since we have often previously spoken in such an intuitive concrete way of the "ψ-vibrations" as though of something quite real. But there is something tangibly real behind the present conception also, namely, the very real electrodynamically effective fluctuations of the electric space-density. The ψ-function is to do no more and no less than permit of the totality of these fluctuations being mastered and surveyed mathematically by a single partial differential equation. We have repeatedly called attention[17] to the fact that the ψ-function itself cannot and may not be interpreted directly in terms of three-dimensional space – however much the one-electron problem tends to mislead us on this point because it is in general a function in configuration space, not real space.

Concerning such a weight-function in the above sense, we would wish its integral over the whole configuration space to remain constantly normalised to the same unchanging value, preferably to unity. We can easily verify that this is necessary if the total charge of the system is to remain constant on the above definitions. Even for non-conservative systems, this condition must obviously be postulated. For, naturally, the charge of a system is not to be altered when, e.g., a light wave falls on it, continues for a certain length of time, and then ceases. (N.B. – This is also valid for ionisation processes. A disrupted particle is still to be included in the system, until

---

[17]End of Part II. (p. 39); paper on Heisenberg's quantum mechanics (p. 60).

the separation is also *logically* – by decomposition of configuration space – completed.)

The question now arises as to whether the postulated *persistence of normalisation* is actually guaranteed by equations (4″), to which ψ is subject. If this were not the case, our whole conception would practically break down. Fortunately, it is the case. Let us form

$$\frac{d}{dt} \int \psi \bar{\psi} \rho \, dx = \int \left(\psi \frac{\partial \bar{\psi}}{\partial t} + \bar{\psi} \frac{\partial \psi}{\partial t} \right) \rho \, dx. \qquad (37)$$

Now, ψ satisfies one of the two equations (4″), and $\bar{\psi}$ the other. Therefore, apart from a multiplicative constant, this integral becomes

$$\int \left(\psi \nabla^2 \bar{\psi} - \bar{\psi} \nabla^2 \psi \right) \rho \, dx = 2i \int \left(J \nabla^2 R - R \nabla^2 J \right) \rho \, dx, \qquad (38)$$

where for the moment we put

$$\psi = R + iJ.$$

According to Green's theorem, integral (38) vanishes identically; the sole necessary condition that functions R and J must satisfy for this – vanishing in sufficient degree at infinity – means physically nothing more than that the system under consideration should practically be confined to a *finite* region.

We can put this in a somewhat different way, by not immediately integrating over the whole configuration space, but by merely changing the time-derivative of the weight-function into a divergence by Green's transformation. Through this we get an insight into the question of the flow of the weight-function, and thus of electricity. The two equations

$$\frac{\partial \psi}{\partial t} = \frac{h}{4\pi i} \left(\nabla^2 - \frac{8\pi^2}{h^2} V \right) \psi$$

$$-\frac{\partial \bar{\psi}}{\partial t} = \frac{h}{4\pi i} \left(\nabla^2 - \frac{8\pi^2}{h^2} V \right) \bar{\psi} \qquad (4'')$$

are multiplied by $\rho\bar{\psi}$ and $\rho\psi$ respectively, and added. Hence

$$\frac{\partial}{\partial t}\left(\rho\psi\bar{\psi}\right) = \frac{h}{4\pi i}\rho\left(\bar{\psi}\nabla^2\psi - \psi\nabla^2\bar{\psi}\right). \qquad (39)$$

To carry out *in extenso* the transformation of the right-hand side, we must remember the explicit form of our many-dimensional, non-Euclidean, Laplacian operator:[18]

$$\rho\nabla^2 = \sum_k \frac{\partial}{\partial q_k}\left[\rho\, T_{p_k}\left(q_l, \frac{\partial\psi}{\partial q_l}\right)\right]. \qquad (40)$$

By a small transformation we readily obtain

$$\frac{\partial}{\partial t}\left(\rho\psi\bar{\psi}\right) = \\ \frac{h}{4\pi i}\sum_k \frac{\partial}{\partial q_k}\left[\rho\,\bar{\psi}T_{p_k}\left(q_l, \frac{\partial\psi}{\partial q_l}\right) - \rho\,\psi T_{p_k}\left(q_l, \frac{\partial\bar{\psi}}{\partial q_l}\right)\right]. \qquad (41)$$

The right-hand side appears as the divergence of a many-dimensional real vector, which is evidently to be interpreted as the *current density of the weight-function* in configuration space. Equation (41) is the *continuity equation* of the weight-function.

From it we can obtain the *equation of continuity of electricity*, and, indeed, a separate equation of this sort is valid for the charge density "originating from each separate particle." Let us fix on the αth particle, say. Let its "charge" be e_α, its mass m_α, and let its coordinate space be described by Cartesians x_α, y_α, z_α, for the sake of simplicity. We denote

---

[18] Cf. paper on Heisenberg's theory, equation (31). The quantity there denoted by $\Delta_p^{-\frac{1}{2}}$ is our "density function" $\rho(x)$ (e.g. $r^2\sin\theta$ in spherical polars). T is the kinetic energy as function of the position coordinates and *momenta*, the suffix at T denoting differentiation with respect to a momentum. In equations (31) and (32), *loc. cit.*, unfortunately by error the suffix k is used twice, once for the summation and then also as a representative suffix in the argument of the functions.

the product of the differentials of the remaining coordinates shortly by dx'. Over the latter, we integrate equation (41), keeping x_α, y_α, z_α, *fixed*. As the result, all terms except three disappear from the right-hand side, and we obtain

$$
\begin{cases}
\dfrac{\partial}{\partial t}\left[e_\alpha \int \psi\bar{\psi}\,dx'\right] = \dfrac{he_\alpha}{4\pi i m_\alpha}\left\{\dfrac{\partial}{\partial x_\alpha}\left[\int\left(\bar{\psi}\dfrac{\partial\psi}{\partial x_\alpha} - \psi\dfrac{\partial\bar{\psi}}{\partial x_\alpha}\right)dx'\right]\right. \\[2mm]
\left. + \dfrac{\partial}{\partial y_\alpha}\left[\int\left(\bar{\psi}\dfrac{\partial\psi}{\partial y_\alpha} - \psi\dfrac{\partial\bar{\psi}}{\partial y_\alpha}\right)dx'\right] + \cdots\right\} \\[2mm]
= \dfrac{he_\alpha}{4\pi i m_\alpha}\,\operatorname{div}_\alpha\left[\int\left(\bar{\psi}\operatorname{grad}_\alpha\psi - \psi\operatorname{grad}_\alpha\bar{\psi}\right)dx'\right].
\end{cases}
$$

$$(42)$$

In this equation, div and grad have the usual three-dimensional Euclidean meaning, and x_α, y_α, z_α are to be interpreted as Cartesian coordinates of real space. The equation is the continuity equation of *that* charge density which "originates from the αth particle". If we form all the others in an analogous fashion, and add them together, we obtain the total equation of continuity. Of course, we must emphasize that the interpretation of the integrals on the right-hand side as *components of the current density*, is, as in all such cases, not absolutely compulsory, because a divergence-free vector could be added thereto.

To give an example, in the conservative *one*-electron problem, if ψ is given by

$$\psi = \sum_k c_k u_k e^{2\pi i \nu_k t + i\theta_k} \quad (c_k,\ \theta_k \text{ real constants}), \qquad (43)$$

we get for the *current density* J

$$J = \frac{he_1}{2\pi m_1}\sum_{(k,l)} c_k c_l(u_l \operatorname{grad} u_k - u_k \operatorname{grad} u_l)$$
$$\times \sin\left[2\pi(\nu_k - \nu_l)l + \theta_k - \theta_l\right]. \qquad (44)$$

We see, and this is valid for conservative systems generally, that, if only a single proper vibration is excited, the current components disappear and the distribution of electricity is constant in time. The latter is also immediately evident from the fact that $\psi\bar{\psi}$ becomes constant with respect to the time. This is still the case even when several proper vibrations are excited, if they all belong to the same proper value. On the other hand, the current density then no longer needs to vanish, but there may be present, and generally is, a *stationary* current distribution. Since the one or the other occurs in the unperturbed normal state at any rate, we may in a certain sense speak of a *return to electrostatic and magnetostatic atomic models*. In this way the lack of radiation in the normal state would, indeed, find a startingly simple explanation.

I hope and believe that the present statements will prove useful in the elucidation of the magnetic properties of atoms and molecules, and further for explaining the flow of electricity in solid bodies.

Meantime, there is no doubt a certain crudeness in the use of a *complex* wave function. If it were unavoidable *in principle*, and not merely a facilitation of the calculation, this would mean that there are in principle *two* wave functions, which must be used *together* in order to obtain information on the state of the system. This somewhat unacceptable inference admits, I believe, of the very much more congenial interpretation that the state of the system is given by a real function and its time-derivative. Our inability to give more accurate information about this is intimately connected with the fact that, in the pair of equations (4″), we have before us only the *substitute* – extraordinarily convenient for the calculation, to be sure – for a real wave equation of probably the fourth order, which, however, I have not succeeded in forming for the non-conservative case.

Zürich, Physical Institute of the University.
(Received June 23, 1926.)

THE COMPTON EFFECT

(*Annalen der Physik* (4), vol. 82, 1927)

It is well known that according to the wave theory of light all changes in the *frequency* and in the *wave-normal* can be predicted by means of very simple and general considerations with respect to *phase*, without going into any details of the process. I mean considerations of the following kind:

Let the xy-plane be the surface of separation of two media with refractive indices n (for $z > 0$) and n' (for $z < 0$), and let a wave of light with the phase

$$2\pi\nu \left[t - \frac{n}{c}(\alpha x + \beta y + \gamma z)\right]$$

fall on it from the positive z-direction. If we assign to the refracted wave the phase

$$2\pi\nu' \left[t - \frac{n'}{c}(\alpha' x + \beta' y + \gamma' z) + \delta\right],$$

and stipulate that for $z = 0$ the phase difference between the waves is to be constant , i.e. independent of x, y, t, we obtain

$$\nu' = \nu, \quad n'\alpha' = n\alpha, \quad n'\beta' = n\beta,$$

i.e. Snell's law of refraction. The reasoning is so general that it holds without alteration, e.g. for crystals, and is immediately transferable to the case of a moving surface of separation. A more detailed investigation of the electromagnetic process only becomes necessary if we are also concerned with the *intensities* (Fresnel's formulae of reflection).

If now we are right in supposing that in de Broglie's waves we have at our disposal a means (ranking on a level with wave optics) for mastering those processes which have previously been thought of exclusively as motions of corpuscles, it is to be expected, nay, even demanded, that we should be able, by means of quite simple phase considerations of the kind mentioned above, to explain the connection between the changes in direction and frequency of the ether wave which occur in the Compton effect and the change of velocity of the electron. For according to de Broglie's idea the latter also can be described as the change of direction and frequency of a wave, namely, of a de Broglie wave. A more detailed investigation of the wave mechanics of the process, such as W. Gordon[1] has recently carried out with complete success, is unnecessary, except for the determination of the intensities. As this is rather lengthy and involved, the simple intuitive treatment given below, which gives everything *but* the intensity, may be welcome.

We start from a result of classical optics. "If in a transparent, homogeneous, and isotropic medium, the refractive index of which depends on the density, a ray of light of wave-length λ crosses a wave of compression (sound wave) of wave-length Λ, then (as L. Brillouin[2] has shown by purely classical reasoning) the ray of light is in part regularly reflected from the planes of the sound wave, provided that the two wave-lengths and the glancing angle θ are connected by Bragg's relation (well known in the theory of X-ray reflection)

$$2\Lambda \sin \theta = \lambda \tag{1}$$

---

[1] W. Gordon, *Ztschr. f. Phys.* 40, p. 117, 1926. Herr Gordon kindly let me see the manuscript of his paper, and hence I was led to the following simple way of looking at the matter, which is, in a nutshell, the basis of Herr Gordon's own treatment.

[2] L. Brillouin, *Ann. d. Phys.* 17, p. 88, 1923.

for *first*-order reflection $(= \lambda$, not $= k\lambda)$. This approximation holds good, provided we can regard the velocity of light as very great compared to the velocity of sound. More accurately, the circumstances are the same as in the case of a *moving* mirror: the angle of reflection is not exactly equal to the angle of incidence, the ray of light undergoes a Doppler displacement, and equation (1) also must be corrected as in the case of a *moving* crystal."

These sentences are taken from an earlier paper,[3] in which it is satisfactorily proved that Brillouin's result can also be obtained on the hypothesis that exchanges of energy and momentum between the two waves proceed by quanta. At that time we were all of the opinion that our whole interpretation of Nature must ultimately rest on such quantum balances, and we rejoiced whenever a trustworthy classical result could be transferred from the old foundation to the new without any trouble. Now we are going in the reverse direction, so to speak. We show that the wave mechanics can provide for the Compton relationships an interpretation which is closely related to the above-mentioned result of Brillouin's, and which is just as simple as the quantum considerations of momentum-energy.

A plane sine wave

$$\psi \sim e^{\frac{2\pi i}{h}\left[h\nu t - \frac{h\sqrt{\nu^2 - \nu_0^2}}{c}(\alpha x + \beta y + \gamma z)\right]},\qquad (2)$$

where

$$\alpha^2 + \beta^2 + \gamma^2 = 1, \quad \nu_0 = \frac{m_0 c^2}{h},$$

(m_0 = rest-mass of the electron, h = Planck's constant, c = velocity of light)

---

[3] E. Schrödinger, *Physik. Ztschr.* 25, p. 89, 1924.

198

for space with no field, is a solution of the relativistic ψ-wave equation which has recently been proposed on many sides,[4]

$$\nabla^2\psi - \frac{1}{c^2}\ddot{\psi} - \frac{4\pi^2\nu_0^2}{c^2}\psi = 0.$$

According to de Broglie, the above solution belongs to an electron moving with energy $h\nu$ in the direction α, β, γ. Hence we deduce in a well-known way that

$$\frac{h\nu}{c}, \quad \frac{h\sqrt{\nu^2 - \nu_0^2}}{c}\alpha, \quad \frac{h\sqrt{\nu^2 - \nu_0^2}}{c}\beta, \quad \frac{h\sqrt{\nu^2 - \nu_0^2}}{c}\gamma,$$

is the four-vector "energy-momentum" of the corresponding electron. From the wave standpoint we will call it the four-vector "propagation," and we shall apply this expression to the coefficients of $ct, -x, -y, -z$, in the phase (dropping the factor $2\pi/h$) *of a completely arbitrary plane sine wave*, be it a ψ-wave, an ether wave, or any other. "Propagation" is therefore a purely wave-kinematical idea; its components are

$$\frac{h\nu}{c} \cdot \text{frequency}, \quad \frac{h\alpha}{\text{wave length}}, \quad \frac{h\beta}{\text{wave length}}, \quad \frac{h\gamma}{\text{wave length}}, \quad (3)$$

where α, β, γ are the direction cosines of the normal to the wave. For an ether wave these quantities coincide with the quantum theory values of energy and momentum. These references to quantum magnitudes, however, are only for the purpose of making it subsequently easier to identify our results with Compton's: we shall work with the purely wave-kinematic idea of propagation (3). By the *three*-vector propagation, we mean, of course, the projection in space, i.e. the vector (3) with the first component omitted.

---

[4]O. Klein, *Ztschr. f. Phys.* 37, p. 895, 1926 ; E. Schrödinger, *Ann. d. Phys.* 81, p. 109, 1926; V. Fork, *Ztschr. f. Phys.* 38, p. 242, 1926 ; Th. de Donder and H. van den Dungen, *Compt. rend.*, July 5, 1926; L. de Broglie, *Compt. rend.*, July 26, 1926; J. Kudar, *Ann. d. Phys.* 81, p. 632, 1926; W. Gordon, *loc. cit.*

According to the hypothesis of wave mechanics, which up to now has always proved trustworthy, it is not the ψ-function itself but the square of its absolute value that is given a physical meaning, namely, density of electricity.[5] A single ψ-wave of type (2) therefore produces a density distribution which is constant both in space and time. If, however, we superpose two such waves, the constants of the second being α', β', γ', we easily see that a "wave of electrical density" arises from their combination, with a propagation vector which is the vector difference of the propagation vectors of the two constituent ψ-waves. If we denote the latter vectors symbolically by A, A', that of the density wave is[6]

$$D = A - A'. \qquad (4)$$

Now it is this density wave that takes the place of the sound wave of Brillouin's paper. If we assume that a light wave is reflected from it as from a moving mirror, subject, however, to fulfilment of Bragg's law, then, as we shall show, our four waves, namely, the two ψ-waves A and A' and the incident and reflected light waves, stand exactly in the Compton relationship. The difference between this case and Brillouin's case of reflection from a sound wave is quantitative only, inasmuch as the velocity of our density wave D is not in general small compared with the velocity of light; on the contrary, arbitrary values up to the velocity of light may occur (but, as is easily verified, never greater than the velocity of light).

The proof of our assertion is easily produced. In fact we do not really require to investigate reflection at a *moving* mirror. As all the four waves, and of course their propagation vectors also, are invariant with respect to Lorentz transformation, we can bring the density wave to rest by means of such a

---

[5] The relativistic refinement of this statement (W. Gordon, *loc. cit.*) does not affect our case at all.

[6] The sign is of small importance, as changing it merely causes the ψ-waves to change places.

transformation. The first component (the time component) of its propagation vector then vanishes. Moreover, the frequency (and wave length) of the light wave are *not* changed during the reflection in this case, i.e. the time component of *its* propagation vector remains unaltered on reflection. Finally, Bragg's relationship holds exactly in the form (1), if λ denotes the wave length of the light wave, Λ that of the density wave, and θ the glancing angle. It can be put in the form

$$2\frac{h}{\lambda}\sin\theta = \frac{h}{\Lambda}, \tag{5}$$

which is illustrated in the accompanying figure (Fig. 1), in which the equality of the angles of incidence and reflection is also taken into account.

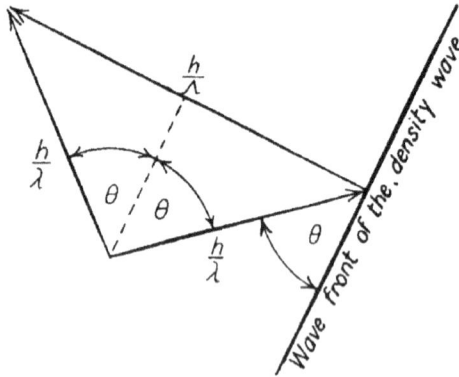

Equation (5) therefore expresses the fact that the three-vector of the incident light wave and the three-vector of D are together equal to the three-vector of the reflected light wave. Moreover, according to what has been said above, a similar relationship holds for the time components also: for D, the time component is *zero*, and for the light wave it is *unaltered* on reflection. If we call the propagation four-vectors for the incident and reflected light waves L and L' respectively, we can sum up all this in the single four-vector equation[7]

$$L + D = L', \qquad (6)$$

which must now hold for an arbitrary four-dimensional system of coordinates. Combined with (4), this equation gives

$$L + A = L' + A'. \qquad (7)$$

If we bear in mind the significance of the components of L, L' from the light quantum point of view, and of those of A, A' according to de Broglie's correlation of ψ-waves and electrons, equation (7) agrees exactly with the statement of Compton's energy-momentum theory of the Compton effect.

It is quite interesting to notice the *complete reciprocity* between the ψ-waves on the one hand and the light waves on the other. The phenomenon may equally well be regarded as a Bragg's reflection of a ψ-wave at the system of *interference fringes* produced by two light waves crossing one another. In the special system of coordinates used above, this system is *at rest* and is identical with O. Wiener's system of stationary light waves. The relationships (4) and (6) express the fact that the system of interference fringes and the density wave coincide, both having the propagation vector D. The special system of coordinates is just the one which W. Pauli[8] formerly found the most convenient for the investigation of the Compton effect.

Fig. 2 is an attempt to represent the relationships between the four wave fronts penetrating through each other and the common stationary wave (in dotted lines) in the special space-time system.[9] To avoid confusing the diagram, the two light

---

[7]The sign of D in (6) is of small importance, as changing it merely causes the light waves to change places.

[8]W. Pauli, jun., *Ztschr. f. Phys* . 18, p. 272, 1923.

[9]Note added to the English edition:

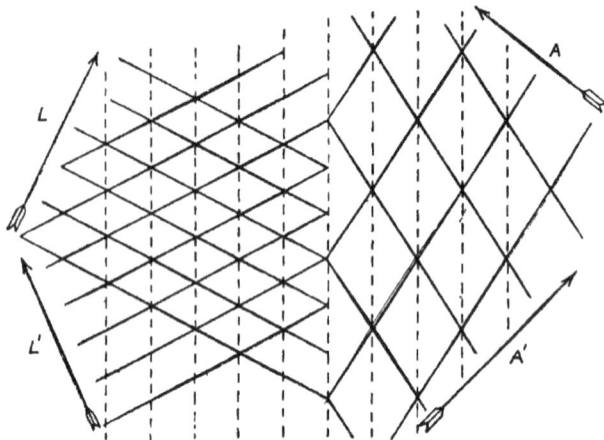

FIG. 2.

wave fronts are drawn in the left half only, and the two ψ-wave fronts in the right half only. The *arrows* are normal to the wave fronts and indicate their directions of propagation. Their length has no meaning. The reader should imagine them pushed together into the middle of the figure by parallel translation, so that the feathers of L' and A' and the heads of L and A all coincide in one point. It is easily deduced from the figure that Bragg's equation (1) holds for either pair of waves (L, L') and (A, A'), if we regard the stationary wave as the "crystal." We may therefore say:

The direction and frequency laws of the Compton effect are completely equivalent to the statement that the pair of light waves and the pair of ψ-waves concerned stand in the Bragg relationship for reflection of the first order (generalised for a moving crystal) to one and the same "space lattice". The

---

Professor Ehrenfest has drawn my attention to the fact that unfortu-nately Fig. 2 is *wrong*. The pair of waves (A, A') ought to be equal to (L', L) in *every* respect, i.e. in wave length and direction also – the wave normal of A' being parallel to that of L and the wave normal of A to that of L'.

*space lattice considered can at the outset have an arbitrary
position and an arbitrary distance between the plane layers,
and a velocity of translation less than that of light but otherwise arbitrary.*

I should now like to deal with one objection to the principle of the method. It might be said: yes, but in the Compton effect the original data are *one* light wave, and *one* electron moving in a specified way, i.e. *one* ψ-wave, as we say:
now wherever does the *second* suitably chosen ψ-wave come
from, which, together with the given one, is to form a suitable "Bragg's mirror" for the given light wave? The reply is
that such simple considerations of phase as we have employed
here are of course absolutely inadequate for the answering of
such questions. By means of these simple considerations we
investigate the Compton phenomenon *in a steady state*, so to
speak, in which the primary wave of one kind is continually
transformed into a secondary wave by reflection at the system
of interference fringes of the other kind, and *vice versa*. This
is just how we proceed in analogous discussions in optics, even
when we study the subject much more accurately by means of
a detailed theory. There also we do not in general study the
first appearance, e.g. of a reflected and a refracted wave at
the head of the primary wave train, but we make an *assumption* not merely for the incident wave alone but likewise for
all the other waves whose appearance can be foreseen; and by
this assumption we seek to represent a stationary state which
will satisfy all the conditions of the problem.

Zürich, Physical Institute of the University.
(Received November 30, 1926.)

THE ENERGY-MOMENTUM THEOREM FOR MATERIAL WAVES

(*Annalen der Physik* (4), vol. 82, 1927)

The *Hamiltonian principle*, from which the exact relativistic differential equation for de Broglie's waves can be obtained,[1] appears to justify completely the hopes which I had set upon an intimate blending together of wave mechanics and classical electrodynamics.[2] If to the integrand (the "Lagrangian function") we add $\mathfrak{H}^2 - \mathfrak{E}^2$, the well-known Lagrangian function of the electromagnetic field when no charges are present, then by varying the potentials as well as the ψ-function we obtain simultaneously the four wave equations for the former with the components of the four-current on the right-hand sides, i.e. the complete electrodynamics. This is due to the fact, first noticed by Gordon, *loc. cit.*, that differentiation of the Lagrangian function for de Broglie's waves with respect to a component of potential gives the corresponding component of the four-current. A most important further result is the *energy-momentum theorem* for the whole field, from which the contribution of the charges, i.e. of the ψ-function, to the energy-momentum tensor may easily be deduced. It is quite clear to me that all this must somehow be

---

[1] O. Klein, *Ztschr. f. Phys.* 37, p. 895, 1926; V. Fock, *ibid.* 38, p. 242, 1926 ; J. Kudar, *Ann. d. Phys.* 81, p. 632, 1926; W. Gordon, *Ztschr. f. Phys.* 40, p. 117, 1926.

[2] *Ann. d. Phys.* 79, p. 754, 1926.

206

included in the very general theories proposed by O. Klein[3] and de Donder.[4] It does not seem superfluous, however, to set forth these connections in as simple a form as possible without referring to the theory of gravitation or the interesting fifth dimension, especially if we bear in mind the very considerable gulf which still yawns between experiment and this beautiful and self-contained theory of the field (see the end of this paper).

We apply the wave equation and Hamilton's principle in the form given by Gordon (*loc. cit.*). The former (always summing from 1 to 4 for indices which occur twice) is as follows:

$$\left[\left(\frac{\partial}{\partial x_\alpha} + i\phi_\alpha\right)\left(\frac{\partial}{\partial x_\alpha} + i\phi_\alpha\right) - k^2\right]\psi = 0, \qquad (1)$$

where

$$x_1, x_2, x_3 = x, y, z; \quad x_4 = ict;$$

$$\phi_1, \phi_2, \phi_3 = \frac{2\pi e}{hc} \cdot \mathfrak{U}_x, \mathfrak{U}_y, \mathfrak{U}_z; \quad \phi_4 = \frac{2\pi e}{hc} \cdot iV.$$

$$k^2 = \frac{4\pi^2 m_0^2 c^2}{h^2}. \qquad (2)$$

$\mathfrak{U}, V$, are the potentials; e, m_0, c, h, the familiar universal constants; $i = \sqrt{-1}$. It must specially be emphasized that the relationship to actual reality is *not* obliterated by the introduction of four-vectors with an imaginary fourth component. This procedure is merely a formal device of calculation, used in order that we may not be obliged to insert the fourth term specially in all four-sums on account of its different sign. The passage to the conjugate complex quantity therefore only affects i when it occurs explicitly, and the ψ-function.

---

[3] *Loc. cit.*
[4] Th. de Donder and H. van den Dungen, *Compt. rend.*, July 5, 1926.

According to Gordon (*loc. cit.*), (1) can be derived from a four-dimensional Hamiltonian integral with the (real) Lagrangian function

$$L_m = (\psi_\alpha + i\phi_\alpha \psi)(\bar{\psi}_\alpha + i\phi_\alpha \bar{\psi}) + k^2 \psi\bar{\psi}, \qquad (3)$$

where the bar denotes the conjugate complex quantity, and we have put for brevity

$$\psi_\alpha = \frac{\partial \psi}{\partial x_\alpha}, \quad \bar{\psi}_\alpha = \frac{\partial \bar{\psi}}{\partial x_\alpha}. \qquad (4)$$

The index α must therefore take effect *after* the bar (see above). In the formation of the variational derivatives, ψ and $\bar{\psi}$ are to be varied *independently*, as Gordon has observed. It is easy to see that this comes to the same thing as varying the real and imaginary parts of ψ independently (which would really be the rational procedure). Hence one of the variational derivatives is

$$\frac{\partial}{\partial x_\alpha}\left(\frac{\partial L_m}{\partial \bar{\psi}_\alpha}\right) - \frac{\partial L_m}{\partial \bar{\psi}_\alpha} = 0, \qquad (5)$$

which is identical with (1); the other gives nothing new. By multiplying (5) by $\bar{\psi}$ we easily obtain

$$\frac{\partial}{\partial x_\alpha}\left(\bar{\psi}\frac{\partial L_m}{\partial \bar{\psi}_\alpha}\right) = \bar{\psi}_\alpha\frac{\partial L_m}{\partial \bar{\psi}_\alpha} + \bar{\psi}\frac{\partial L_m}{\partial \bar{\psi}} = L_m; \qquad (6)$$

the latter equality follows from the fact that L_m is homogeneous and of the first degree in the five quantities $\bar{\psi}\bar{\psi}_\alpha$. The right-hand side is unaltered by passage to the conjugate complex quantities; and hence by subtraction

$$\frac{\partial}{\partial x_\alpha}\left(\bar{\psi}\frac{\partial L_m}{\partial \bar{\psi}_\alpha} - \psi\frac{\partial L_m}{\partial \psi_\alpha}\right) = 0, \qquad (7)$$

This, according to Gordon, is the *equation of continuity for electricity*. We notice that

$$\bar{\psi}\frac{\partial L_m}{\partial \bar{\psi}_\alpha} - \psi\frac{\partial L_m}{\partial \psi_\alpha} = i\frac{\partial L_m}{\partial \phi_\alpha}. \qquad (8)$$

We define the *four-current* as follows,

$$s_\alpha = -\lambda \frac{\partial L_m}{\partial \phi_\alpha} \tag{9}$$

where λ is a universal real constant to be chosen later. By the quantities s a we mean the four quantities which in Lorentz's theory were given by

$$s_1, s_2, s_3 = \rho \frac{\mathfrak{b}}{c}, \quad s_4 = i\rho. \tag{10}$$

We now extend our Lagrangian function (3) in such a way that by varying the quantities ϕ_α in it we obtain (as is possible, in consequence of (9)) the laws of the electromagnetic field. We put

$$L_e = \frac{1}{4} f_{\alpha\beta} f_{\alpha\beta} = \frac{1}{4} \left(\frac{\partial \phi_\beta}{\partial x_\alpha} - \frac{\partial \phi_\alpha}{\partial x_\beta} \right) \left(\frac{\partial \phi_\beta}{\partial x_\alpha} - \frac{\partial \phi_\alpha}{\partial x_\beta} \right), \tag{11}$$

where for brevity

$$f_{\alpha\beta} = \frac{\partial \phi_\beta}{\partial x_\alpha} - \frac{\partial \phi_\alpha}{\partial x_\beta}. \tag{12}$$

From (2) and familiar formulae, we see that the quantities f have the following meanings:

$$
\begin{aligned}
f_{14} &= -\frac{2\pi e}{hc} i\, \mathfrak{E}_x, \quad f_{24} = -\frac{2\pi e}{hc} i\, \mathfrak{E}_y, \quad f_{34} = -\frac{2\pi e}{hc} i\, \mathfrak{E}_z; \\
f_{23} &= \frac{2\pi e}{hc} \mathfrak{H}_x, \quad f_{31} = \frac{2\pi e}{hc} \mathfrak{H}_y, \quad f_{12} = \frac{2\pi e}{hc} \mathfrak{H}_z;
\end{aligned}
\tag{13}
$$

where $\mathfrak{E}$, $\mathfrak{H}$ represent the field in the customary units. As Lagrangian function we now take

$$L = L_m + L_e \tag{14}$$

and obtain in a familiar way

$$\frac{\partial f_{\alpha\beta}}{\partial x_\alpha} = \frac{\partial L}{\partial \phi_\beta} = -\frac{s_\beta}{\lambda} \tag{15}$$

by varying ϕ_β. If the constant λ is given the value

$$\lambda = \frac{hc}{8\pi^2 e},$$

the equations (15) represent the so-called *second quartet of the Maxwell-Lorentz equations*, while the *first* quartet is identically satisfied by (12). Using (12) and Maxwell's subsidiary condition ($\frac{\partial \phi_\alpha}{\partial x_\alpha} = 0$) the equations (15) become the wave equations for the potentials,

$$\frac{\partial^2 \phi_\beta}{\partial x_\alpha \partial x_\alpha} = -\frac{s_\beta}{\lambda}. \tag{15$'$}$$

From (15$'$) (together with Maxwell's subsidiary condition), it is now easy to verify that

$$\frac{\partial T_{\rho\sigma}}{\partial x_\sigma} = -\frac{f_{\rho\sigma} s_\sigma}{\lambda}, \tag{16}$$

where

$$T_{\rho\sigma} = f_{\rho\alpha} f_{\sigma\alpha} - \delta_{\rho\sigma} L_e \tag{17}$$

is Maxwell's well-known stress-energy-momentum-tensor (apart from a universal constant). In Lorentz's work the right-hand side of (16) means the energy or momentum abstracted from the electrons by the forces of the field. In virtue of (9) and the ψ-wave-equation (5), the right-hand side of (16) may likewise be represented as the divergence of a tensor, namely, of the energy-momentum-tensor of the charge (or of the "matter"). In the first place, we have

$$-\frac{f_{\rho\sigma} s_\sigma}{\lambda} = \left(\frac{\partial \phi_\sigma}{\partial x_\rho} - \frac{\partial \phi_\rho}{\partial x_\sigma} \right) \frac{\partial L_m}{\partial \phi_\sigma}$$
$$= \frac{\partial L_m}{\partial \phi_\sigma} \frac{\partial \phi_\sigma}{\partial x_\rho} - \frac{\partial}{\partial x_\sigma} \left(\phi_\sigma \frac{\partial L_m}{\partial \phi_\sigma} \right), \tag{18}$$

where the latter equation follows from the fact that the four-current is free of sources (from equations (7) and (8)). Further, we note that

$$
\begin{aligned}
\frac{\partial L_m}{\partial x_\rho} = {} & \frac{\partial L_m}{\partial \phi_\sigma} \frac{\partial \phi_\sigma}{\partial x_\rho} + \frac{\partial L_m}{\partial \bar{\psi}} \bar{\psi}_\rho + \frac{\partial L_m}{\partial \psi} \psi_\rho \\
& + \frac{\partial L_m}{\partial \bar{\psi}_\sigma} \frac{\partial \bar{\psi}_\sigma}{\partial x_\rho} + \frac{\partial L_m}{\partial \psi_\sigma} \frac{\partial \psi_\sigma}{\partial x_\rho}.
\end{aligned}
\tag{19}
$$

Since by (4), however,

$$
\frac{\partial \psi_\sigma}{\partial x_\rho} = \frac{\partial \psi_\rho}{\partial x_\sigma}, \text{ etc.,}
\tag{20}
$$

it is possible to transform the last two terms in (19) as is done in integration by parts; after this transformation four terms cancel, on account of (5), and we obtain

$$
\frac{\partial L_m}{\partial x_\rho} = \frac{\partial L_m}{\partial \phi_\sigma} \frac{\partial \phi_\sigma}{\partial x_\rho} + \frac{\partial}{\partial x_\sigma} \left(\bar{\psi}_\rho \frac{\partial L_m}{\partial \bar{\psi}_\sigma} + \psi_\rho \frac{\partial L_m}{\partial \psi_\sigma} \right).
\tag{21}
$$

We subtract this from equation (18), and we have

$$
\left\{
\begin{aligned}
-\frac{f_{\rho\sigma} s_\sigma}{\lambda} &= \frac{\partial L_m}{\partial x_\rho} - \frac{\partial}{\partial x_\sigma} \left(\bar{\psi}_\rho \frac{\partial L_m}{\partial \bar{\psi}_\sigma} + \psi_\rho \frac{\partial L_m}{\partial \psi_\sigma} + \phi_\rho \frac{\partial L_m}{\partial \phi_\sigma} \right) \\
&= \frac{\partial}{\partial x_\sigma} \left(\delta_{\rho\sigma} L_m - \bar{\psi}_\rho \frac{\partial L_m}{\partial \bar{\psi}_\sigma} - \psi_\rho \frac{\partial L_m}{\partial \psi_\sigma} - \phi_\rho \frac{\partial L_m}{\partial \phi_\sigma} \right) \\
&= \frac{\partial S_{\rho\sigma}}{\partial x_\sigma},
\end{aligned}
\right.
\tag{22}
$$

where we introduce the *energy-tensor of the charges or of the "matter"*

$$
S_{\rho\sigma} = \bar{\psi}_\rho \frac{\partial L_m}{\partial \bar{\psi}_\sigma} + \psi_\rho \frac{\partial L_m}{\partial \psi_\sigma} + \phi_\rho \frac{\partial L_m}{\partial \phi_\sigma} - \delta_{\rho\sigma} L_m.
\tag{23}
$$

From (16) and (22) we obtain

$$
\frac{\partial}{\partial x_\sigma} (T_{\rho\sigma} + S_{\rho\sigma}) = 0
\tag{24}
$$

for the combined laws of conservation of energy and momentum for the electromagnetic field and de Broglie's wave-field taken together.

Calculation shows that the tensor $S_{\rho\sigma}$ is *symmetrical*. We easily find the explicit expression

$$\begin{cases} S_{\rho\sigma} = \bar{\psi}_\rho \psi_\sigma + \bar{\psi}_\sigma \psi_\rho + i\phi_\sigma(\bar{\psi}_\rho \psi - \psi_\rho \bar{\psi}) \\ +i\phi_\rho(\bar{\psi}_\sigma \psi - \psi_\sigma \bar{\psi}) + 2\psi\bar{\psi}\phi_\rho\phi_\sigma - \delta_{\rho\sigma}L_m, \end{cases} \quad (25)$$

or the following one, which is more closely related to the form of L_m given in (3),

$$\begin{cases} S_{\rho\sigma} = (\psi_\rho + i\phi_\rho\psi)(\bar{\psi}_\sigma - i\phi_\sigma\bar{\psi}) \\ +(\psi_\sigma + i\phi_\sigma\psi)(\bar{\psi}_\rho - i\phi_\rho\bar{\psi}) - \delta_{\rho\sigma}L_m. \end{cases} \quad (25')$$

In contradistinction to $T_{\sigma\sigma}$, the Laue scalar (diagonal sum) $S_{\sigma\sigma}$ does *not* vanish. We easily obtain

$$S_{\sigma\sigma} = -2(L_m + k^2\psi\bar{\psi}). \quad (26)$$

The complete tensor can also be represented by means of the complete Lagrangian function in the following way, well known from similar cases:

$$\begin{cases} T_{\rho\sigma} + S_{\rho\sigma} &= \dfrac{\partial L}{\partial\left(\frac{\partial\phi_\rho}{\partial x_\alpha}\right)}\dfrac{\partial\phi_\sigma}{\partial x_\alpha} + \dfrac{\partial L}{\partial\left(\frac{\partial\phi_\alpha}{\partial x_\rho}\right)}\dfrac{\partial\phi_\alpha}{\partial x_\sigma} + \dfrac{\partial L}{\partial\bar{\psi}_\rho}\bar{\psi}_\sigma \\ &+ \dfrac{\partial L}{\partial\psi_\rho}\psi_\sigma + \dfrac{\partial L}{\partial\phi_\rho}\phi_\sigma - \delta_{\rho\sigma}L. \end{cases} \quad (27)$$

This is analogous to the representation of the Hamiltonian function by means of the Lagrangian function in *point* mechanics.

It should be remembered that our tensor components $S_{\rho\sigma}$ and $T_{\rho\sigma}$ have the physical dimensions cm.$^{-4}$. Before being applied they must be multiplied by the constant

$$\frac{h^2 c^2}{32\pi^3 e^2},$$

which is of the dimensions of the square of a charge, in order that they may physically represent energy, momentum, and stress. (N.B. Other defects in the dimensions are to be rectified, as is known, by means of powers of c.)

If we now ask ourselves whether this self-contained theory of the field – apart from the provisional neglect of the electronic spin – corresponds to reality in the way we had previously hoped for from such theories, the question must be answered *in the negative*. The examples worked out, particularly that of the H-atom, show in fact that we have *not* to insert in the wave equation (1) those potentials which result from the potential equations (15′) with the four-current (9). On the contrary, we know that in the case of the H-atom we have to substitute the *given* potentials of the nucleus and of possible "external" electromagnetic fields for the ϕ_α's in (1), and solve the equation for ψ. The distribution of current produced by this ψ is then calculated from (9), and from the distribution the potentials produced by it are found by (15′). By adding the latter to the potentials given in advance, we obtain those potentials which define the external action of the atom as a whole. We thus obtain (with a suitable normalisation of ψ, for which, it must be admitted, a proof by the theory of the field is also wanting) the *neutralisation* of the nuclear charge at greater distances on the one hand, and on the other hand the *radiation*. With reference to the attempt, which it would now be natural to make, to substitute these newly found potentials in equation (1), and thus to calculate a "second approximation", it is to be remarked that we must not on any account proceed in this way with the *neutralisation potential*, as it would *completely* alter the values of the terms, and hence would make many more stages of approximation necessary. These, even if the process converges at all, certainly do *not* lead to the correct hydrogen terms, much less (in the case of nuclear charge 2) to the helium *atom* terms. On the contrary, we should very probably obtain the required *radiation correction*[5] by dealing with the radiation potentials in the way described, if we suppose that *one* proper vibration is *strongly* excited but all the others only very feebly.

Hence there is something which intrudes into the self-contained system of field equations in a peculiar way. This is not yet fully intelligible at present, but it must be considered in connection with the two following facts:

1. The exchange of energy and momentum between the electromagnetic field and "matter" does *not* in reality take place continuously as the expression (24) in terms of the field would lead us to believe.

2. In Lorentz's theory also we have to substitute in the first instance only the fields of the *other* electrons in the equations of motion of the single electron, and not its own individual field. The reaction of the latter has *already* been almost entirely taken account of as electromagnetic *mass, in setting up the equations of motion.* The corresponding term in equation (1) is the term with k^2. The reaction of radiation results in a second appproximation from the reaction of the electron's own field in Lorentz's theory also.

The question whether the solution of the difficulty is really to be found only in the purely *statistical* interpretation of the field theory which has been proposed in several quarters[6] must for the present be left unsettled. Personally I no longer regard this interpretation as a finally satisfactory one,[7] even if it proves useful in practice. To me it seems to mean a renunciation, much too fundamental in principle, of all attempt to understand the individual process.

---

[5]Cf. *Ann. d. Phys.* 81, p. 129 *et seq.*, 1926 (p. 116 *et seq.*).

[6]M. Born, *Ztschr. f. Phys.* 38, p. 803, 1926; 40, p. 167, 1926; P. A. M. Dirac, *Proc. Roy. Soc.*, A, 112, p. 661, 1926; and W. Gordon, *loc. cit.*

[7]Cf. *Die Naturwissenschaften*, 12, p. 720, 1924.

A *brighter* side of the difficulty in question deserves to be mentioned. By interrupting the completeness of the system of field equations in her actual behaviour, Nature accommodates herself to our mathematical powers to an astonishing extent. Even the theory of the hydrogen atom would become immeasurably complicated from the mathematical point of view, if the ϕ_α's did not stand for *given* potential values in equation (1), but if instead we had to add to them those which are to be calculated by means of (9) and (15') from ψ, which is itself unknown.

Zürich, Physical Institute of the University.
(Received December 10, 1926.)

The Exchange of Energy According to Wave Mechanics

(*Annalen der Physik* (4), vol. 83, 1927)

A series of papers which have appeared in this journal[1] forms the starting-point of the present note. Here, in fact, we shall apply the *many*-dimensional form of "wave mechanics" to which these papers are almost exclusively devoted, and which can be brought into coincidence with the Heisenberg-Dirac quantum mechanics, instead of that *four-* (or according to O. Klein *five-*) dimensional form,[2] which corresponds to de Broglie's original conception and possibly strikes more closely at the root of the matter, but which is meanwhile only prospective in character, because we do not yet understand how to formulate the problem for *more* than one electron by means of it. I must ask leave to develop afresh here some important matters which have previously been expounded by others (Heisenberg, Dirac, Jordan). For I should like to remain intelligible to those even who have not yet made themselves familiar with the use of the new number-systems (matrices, q-numbers) employed by those writers.[3]

---

[1]"Quantisation as a Problem of Proper Values," Parts I.-IV. above; henceforth referred to as Q. I.-IV.

[2]O. Klein, *Ztschr. f. Phys.* 37, p. 895, 1926; W. Gordon, *ibid.* 40, p. 117, 1926; Q. IV.; E. Schrödinger, *Ann. d. Phys.* 82.

[3]The difficulties which are very generally experienced may be compared with the following. If someone, e.g., in a lecture, began by expounding the old action-at-a-distance theory of electricity in Cartesian

§ 1. The Method of Variation of Constants[4]

More general methods,[5] which are far superior for many purposes, have since been given for the treatment of the perturbation problem solved in Q. III. (§§ 1 and 2). We consider a conservative system; let its wave equation (Q. IV., equation (4″))

$$\nabla^2\psi - \frac{8\pi^2}{h^2}V\psi - \frac{4\pi i}{h}\psi = 0 \tag{1}$$

have the normalised proper solutions

$$\psi_k e^{\frac{2\pi i E_k t}{h}}, \tag{2}$$

where ψ_k depends only on the coordinates of the system.[6] ψ_k therefore satisfies the equation

$$\nabla^2\psi_k + \frac{8\pi^2}{h^2}\left(E_k - V\right)\psi_k = 0 \tag{3}$$

in which the time does not appear. The general solution of (1) is

$$\psi = \sum_k c_k\psi_k e^{\frac{2\pi i E_k t}{h}}, \tag{4}$$

---

coordinates, and then introduced vector analysis for the first time while passing to Maxwell's theory, the listener would have great difficulty in distinguishing between the physically new *matter* and the new *form*. (Similarly, e.g., in P. A. M. Dirac's paper, *Proc. Roy. Soc.*, A, 14, p. 250, § 3, one may easily overlook the fact that here a quite new physical hypothesis has just been introduced, namely, a "successive" or "double" application of the process which Heisenberg calls "passage to matrices" and Dirac "passage to q-numbers", and which I call "passage to wave-mechanics".)

[4]P. A. M. Dirac, *Proc. Roy. Soc.*, A, 112, p. 674, 1926.
[5]Cf. especially M. Born, *Ztschr. f. Phys.* 40, p. 172, 1926.
[6]The wave function ψ must be considered as essentially complex. We assume the functions of the coordinates ψ_k to be real, merely in order to simplify the formulae.

where the c_k's are arbitrary constants, complex in general, which we call the *amplitudes*. (The squares of their absolute values we call for short the squares of the amplitudes.)

We now bring into play a small perturbation, constant with respect to time, by replacing V in (1) by $V + r$, where r is a small function of the coordinates only. We again attempt to make (4) satisfy the equation perturbed in this way, by regarding the amplitudes as slowly varying functions of the time. By substituting (4) in the (perturbed) equation (1) and bearing (3) in mind, we obtain

$$-\frac{8\pi^2}{h^2} r \sum_k c_k \psi_k e^{\frac{2\pi i E_k t}{h}} - \frac{4\pi i}{h} \sum_k \dot{c}_k \psi_k e^{\frac{2\pi i E_k t}{h}} = 0 \qquad (5)$$

for this functional dependence on the time. As necessary and sufficient condition for the vanishing of the left-hand side we employ the condition that it should be orthogonal to every function of the complete orthogonal system ψ_l. We thus obtain the infinite set of equations

$$\dot{c}_l = \frac{2\pi i}{h} \sum_k \epsilon_{kl} c_k e^{\frac{2\pi i (E_k - E_l)t}{h}}, \qquad (6)$$

where

$$\epsilon_{kl} = \int r \psi_k \psi_l dx. \qquad (7)$$

The equations (6) take everything into account.

If now all the proper value differences are large compared with the "elements of the matrix of perturbation" ϵ_{kl}, each c_k ($k \neq l$) can be taken as approximately constant throughout the period of the exponential factor standing beside it; thus all these terms produce only small oscillatory perturbations in c_l. It is only for the sum-term $k = l$ that this does not hold, because then the exponential factor becomes unity. Apart from these small perturbations, we therefore have

$$\dot{c}_l = \frac{2\pi i}{h} \epsilon_{ll} c_l; \quad c_l = c_l^0 e^{\frac{2\pi i E_k t}{h}}. \qquad (8)$$

Hence the *absolute values* of the amplitudes remain altogether unchanged (to this approximation), but their *phases* undergo secular alterations (which can also be interpreted as *perturbations of proper values*; cf. Q. III.).

If, on the contrary, proper value differences which are comparable with, or even small compared to, the perturbation quantities ϵ_{kl} occur in the unperturbed problem, then the amplitudes of all those proper vibrations which belong to such a group of neighbouring proper values are coupled together by the equations (6), even in the approximation hitherto considered, in such a way that only the sum of the squares of the amplitudes remains constant, instead of, as before, the square of a single amplitude. We see this as follows. We shall consider in particular the case of an α-fold proper value. Let c_l be the amplitude of a corresponding proper vibration. On the right-hand side of (6) α exponentials then become equal to unity, and α secular terms remain, in the approximation considered, namely, just those amplitudes which belong to the proper value in question. The same happens in all the α equations (6) in which one of these amplitudes occurs on the left-hand side. We thus obtain for the determination of these amplitudes the finite and self-contained system of equations

$$\dot{c}_l = \frac{2\pi i}{h} \sum_{k=1}^{\alpha} \epsilon_{kl} c_k; \quad l = 1, 2, \ldots, \alpha, \tag{9}$$

where for simplicity we have numbered the α amplitudes in question from 1 to α. Accordingly, in general an exchange takes place between the amplitudes belonging to one and the same proper value, and – in the approximation considered – with such amplitudes only. If we multiply (9) by the conjugate complex quantity c_l^*, take the real parts, and sum for all values of l, the right-hand side vanishes (on account of the symmetry of the ϵ_{kl}'s), i.e.

$$\sum_{k=1}^{\alpha} c_l c_l^* = \text{const.} \tag{10}$$

is an integral of (9). Apart from this, the equations are of course very easy to integrate, since the ϵ_{kl}'s are constant. We are led precisely to the transformation to principal axes given in Q. III., pp. 73-4. The solution is identical with what were there called the "perturbed solution of zero approximation" and the "perturbed proper values of the first approximation".

§ 2. The Explanation according to Wave Mechanics of the Quantum Exchange of Energy

The very simple state of affairs which has just been described now provides, as Heisenberg[7] and Jordan[8] have remarked, the explanation according to wave mechanics of the fact which can quite well be characterised as the empirical basis of the quantum theory, namely, that physical systems apparently influence each other only when they agree in respect of a "difference of level", or nearly so; and that the influence is always exerted on the four critical levels only, and, moreover, always in such a way that one of the two systems is raised to its higher level at the expense of the other, which undergoes an "equivalent" opposite displacement.

Thus, if we have two systems with the wave equations

$$\nabla_1^2 \psi - \frac{8\pi^2}{h^2} V_1 \psi - \frac{4\pi i}{h} \psi = 0 \tag{11}$$
(proper functions ψ_k for E_k)

and

$$\nabla_2^2 \psi - \frac{8\pi^2}{h^2} V_2 \phi - \frac{4\pi i}{h} \phi = 0 \tag{12}$$

[7]W. Heisenberg, *Ztschr. f. Phys.* 38, p. 411, 1920; 40, p. 601, 1920.
[8]P. Jordan, *ibid.* 40, p. 661, 1927.

(proper functions ϕ_l for F_l)

and imagine them united into *one* system ("with vanishing coupling"), the wave equation of the latter will, as is easily seen, be

$$(\nabla_1^2 + \nabla_2^2)\Psi - \frac{8\pi^2}{h^2}(V_1 + V_2)\Psi - \frac{4\pi i}{h}\dot{\Psi} = 0, \qquad (13)$$

with the proper functions $\psi_k \phi_l$ for the proper values $E_k + F_l$. Let us now add a small coupling term r to $V_1 + V_2$, as in § 1. It will then be a question of whether or not the imagined combination has caused new degenerations, or approximate degenerations (i.e. multiple proper values, or proper values close together), to appear. If this does not happen, i.e. if all the proper values $E_k + F_l$ differ sufficiently, the two systems do not influence each other in the first approximation, that considered in § 1. If, however, new degenerations occur in (13), a secular exchange of amplitudes takes place.

For example, let

$$E_k + F_{l'} = E_{k'} + F_l \qquad (14)$$

for four special values k, k', l, l' (this just means that the two systems agree in respect of the proper value difference $E_k - E_{k'} = F_l - F_{l'}$). Then the *two* proper functions

$$\psi_k \phi_{l'} \text{ and } \psi_{k'} \phi_l \qquad (15)$$

correspond to the proper value (14). If their amplitudes are c_1, c_2, an exchange will take place between them in accordance with the equations

$$\begin{cases} \dot{c}_1 = \frac{2\pi i}{h}(\epsilon_{11}c_1 + \epsilon_{12}c_2) \\ \dot{c}_2 = \frac{2\pi i}{h}(\epsilon_{12}c_1 + \epsilon_{22}c_2), \end{cases} \qquad (16)$$

where the constants ϵ_{ik} are defined by appropriate application of equation 7, § 1.

Evidently we now have to interpret, e.g., an increase in the amplitude corresponding to $\psi_k \phi_l$ at the expense of the other one, in the double sense that just as in the one system the amplitude of ψ_k increases at the expense of that of $\psi_{k'}$, so in the other system the amplitude of $\phi_{l'}$ increases at the expense of that of ϕ_l. We can picture the state of affairs as follows: the wave function of the whole system describes at any moment the state of the first system (if we overlook the small coupling and the existence of the second system), and the reverse statement is equally true. To be sure, simple numbers no longer appear as amplitudes, but instead we have linear combinations of the proper functions of the *other* system, i.e. according to the present interpretation, of a completely external system. This, however, does not cause any serious difficulty. In the calculation of any physical quantity relating to the system under consideration, we have simply to integrate over the coordinates of the external system in a way similar to that described already in Q. IV., § 7. Thus we obtain, e.g., the sum of the squares of the amplitudes of all those proper functions of the whole system which contain ϕ_l, for the square of *the amplitude of ϕ_l*.[9]

We thus find that without assuming discrete energy levels and quantum exchange of energy, and even without having to consider any meaning for the proper values other than *frequencies*, we can give a simple explanation of the fact that physical interaction chiefly takes place between *those* systems

---

[9]The inconvenient fact that if we confine ourselves to the present simple method of calculation the external proper functions cannot be got rid of once for all, i.e. that the complex amplitude of ϕ_l in the isolated system cannot be obtained by itself, appears to be inherent in the nature of the case. For it is not possible actually to do away with the coupling, unless another system, namely, the radiation (or the "ether"), is taken into account as well. The Coulomb coupling terms cease to describe the state of affairs correctly long before they have become negligibly small; they would have to be altered by taking the radiation from one system to the other into account.

222

in which, according to the older conception, "the same energy element occurs". As Heisenberg points out, it is a question of a simple resonance phenomenon with beats, similar to the phenomenon of the so-called sympathetic pendulum. Without quantum postulates we have arrived at an effect which is exactly the same *as if* the quantum postulates were in force. This "as-if" situation is not new to us. The spontaneously emitted frequencies are also obtained, as if the proper values were discrete energy levels and Bohr's frequency condition held good.

According to the fundamental principles of research which are commonly regarded as correct, does not the foregoing compel us to exercise the utmost caution with respect to the quantum postulates, even (I almost feel inclined to say) to distrust them – quite apart from their axiomatic unintelligibility? From the psychological point of view it is clear that as soon as the conception of the "terms" as discrete energy levels had been introduced, we were obliged to see a corroboration of that conception in every new exchange phenomenon discovered, even if there is really nothing present in nature beyond the resonance phenomenon just discussed. The reader is *not* to object: oh, but the conception of the terms as energy levels is raised above all doubts by *researches on electronic impact*, if by nothing else: surely you will not doubt that the potential difference fallen through is a measure of the kinetic energy of the single electron? My reply is this: Yes, I do question whether it is not very much more to the point to push the idea of the frequency of the de Broglie wave into the foreground instead of that of the "kinetic energy of the single electron." It is known that in passing through a potential difference these waves undergo just that change of frequency which corresponds to the acquired kinetic energy. Further, the wave equation gives just those deviated ray-paths which are actually observed in the determination of e/m and v.

I cannot help feeling that to admit the quantum postulates

in addition to the resonance phenomenon is to accept *two* explanations for the same thing. But two explanations are like two excuses: one is certainly untrue, and usually both. In the concluding section we shall add another "as-if" situation to the one described here.

§ 3. A Statistical Hypothesis

If in the case of prolonged interaction of two systems we try to obtain an expression for the average distribution of the amplitudes from the equations (9), then, just as in the analogous case in classical mechanics, the attempt will not succeed without a special supplementary hypothesis of a statistical nature. Like the fundamental equations of mechanics, the equations (9) are clearly unaffected by changing the sign of the time; this change can be compensated for by interchanging i and $-i$ (a change of sign of all the phases, corresponding to the change of sign of all the velocities in classical mechanics). This alone shows that there is no "equalising tendency" inherent in the resonance process itself. In fact, calculation shows that the time-averaged values of the squares of the amplitudes in general depend on their initial values. In order to obtain statistical expressions, a hypothesis as to the *a priori* probability of the initial values is therefore necessary. It appears that only *one* hypothesis is possible if it is to satisfy the following requirements:

(1) The hypothesis is to be independent of the *instant of time* for which it is stated; i.e. the probability that given values of the amplitudes shall occur is not to alter with the passage of time, in consequence of the operation of equations (9).

(2) The hypothesis is to be independent of the particular one of the infinite number of completely equivalent orthogonal systems (arising from arbitrary orthogonal substitution among the proper functions corresponding to the same proper

value) for which it is stated (cf. Q. III., p. 70 *et seq.*).

We may easily convince ourselves that in these circumstances the only possible assumption is the following: the probability density in a space in which the real and imaginary parts of the amplitudes are taken as rectangular coordinates is a function only of the sums of the squares of the amplitudes corresponding to the numerically different proper values.

It follows from this assumption that the average values of the squares of the amplitudes corresponding to the *same* proper value are equal, by symmetry; i.e. each partial sum of these quantities is now proportional to the number of terms in the sum. This is the only consequence that we shall employ in what follows, and that only for cases of extremely high degeneracy and only for partial sums with an extremely great number of terms.

We must refrain from the attempt, by means of anything analogous to the quasi-ergodic[10] hypothesis, to set up these average values as correct time-averages. The equations (9) are much too transparent to be satisfied by any such hypothesis (they possess at least α independent holomorphic integrals, namely, the squares of the amplitudes of the "normal vibrations"). The case is quite analogous to that of idealised solids, in which the constancy of the squares of the amplitudes of the normal vibrations really *ought* to prevent us from applying any statistical reasoning.

I should not like to leave unmentioned the fact that the same assumption regarding the squares of the amplitudes of the proper vibrations corresponding to the same proper value was necessary in the case of the *Stark effect*, in order that correct intensity ratios for the fine structure components should be obtained (cf. Q. III., p. 83).

---

[10]Note to English edition. The "ergodic hypothesis" (Boltzmann) is what Maxwell called the "principle of continuity of path".

§ 4. Arbitrary System in a Heat-bath

We return to the considerations of § 2. We will now assume that in the whole system the proper value (14) *only* is excited initially (and therefore permanently). Further, we will now assume that E_k, $E_{k'}$ F_l, $F_{l'}$ the four proper values of the separate systems which come into consideration, and which in § 2 we tacitly assumed to be *simple*, exhibit multiplicities of orders $\alpha_k, \alpha_{k'}, \alpha_l, \alpha_{l'}$. The proper value (14) then becomes $(\alpha_k \alpha_{l'} + \alpha_{k'} \alpha_l)$-fold, for instead of the *two* degenerate proper functions (15) there appear *two groups*, in number $\alpha_k \alpha_{l'}$ and $\alpha_{k'} \alpha_l$ respectively. According to the statistical hypothesis of § 3, the sum of the squares of the amplitudes of the first group bears the ratio

$$\alpha_k \alpha_{l'} : \alpha_{k'} \alpha_l, \tag{17}$$

to the sum of the squares of the amplitudes of the second group. From what was said at the end of § 2, this is also the ratio of the sum of the squares of the amplitudes of all the proper vibrations corresponding to E_k to the sum of the squares of the amplitudes of all the proper vibrations corresponding to $E_{k'}$ in the first system, considered as isolated.

Thus, according to our statistical hypothesis, the interaction with the external system forces a quite definite value upon the initially *undetermined* ratio of the sums of squares of amplitudes corresponding to *different* proper values. This value is given by the "cross"-products of the degrees of degeneracy. (By "cross"-product the following is meant: the "upper" level of the system itself is to be combined with the lower level of the external system, and *vice versa*). For brevity we will from this point call the sum of the squares of the amplitudes corresponding to a proper value the *excitation strength* of that proper value.

We now pass on to a somewhat more complicated case. We shall still, however, abide by the condition that in the whole system only *one* proper value, which we shall call E, is permanently excited. The second system (ϕ_l, F_l), which we will now call the *heat-bath*, is to be an extremely large system with an exceedingly dense proper value spectrum, such that for *every* E_k of the first system, which we will call the *thermometer*, there always exists a proper value of the heat-bath, $F_{l'}$, satisfying the equation

$$F_{l'} = E - E_k, \qquad (18)$$

where $F_{l'}$ is to be multiple, even to a high order.

Hence a quite definite ratio is given perforce to the excitation strengths of all the proper values E_k of the thermometer; in fact, they are proportional to the products

$$\alpha_k \alpha_{l'}. \qquad (19)$$

The ratios of the $\alpha_{l'}$'s, however, can be determined in a very general way. For the question of the multiplicity $\alpha_{l'}$ of the proper value $F_{l'}$ of the heat-bath, i.e. of the number of essentially different proper functions of the heat-bath which correspond to this proper value, is clearly identical with the question: in how many essentially different ways could we dispose of the *energy* $F_{l'}$ in the heat-bath *if* the latter were "energy-quantised"? This, however, is exactly the question which would be raised in connection with the calculation of the *entropy* of the heat-bath according to Planck's quantum statistics; in the latter, the entropy is taken to be equal to k times the logarithm of the number sought (k is Boltzmann's constant). The only difference[11] is that it is *sufficient* if we put the question in terms of a hypothetical period – the result

---

[11] There are also of course the well-known small differences in the special specification of the "energy levels" by the new quantum mechanics as compared with that by the old ("half-integral" quantisation, etc.). Further, we must remark that as regards what writers at present like to call the *kind of statistics* (Bose and Einstein, Fermi, etc.) absolutely nothing is prejudiced by the very general developments of the text. The distinction only appears when we give effect to a Pauli or a Heisenberg

of the enumeration is of course independent of the form of the statement.

That is, we have

$$k \log a_{l'} = S(E - E_k),$$

where the right-hand side denotes the entropy which the heat-bath with energy $E - E_k$ possesses, according to Planck's quantum statistics. By (19), the excitation strengths of the proper values E_k of the thermometer are therefore proportional to the quantities

$$\alpha_k e^{\frac{1}{k}S(E - E_k)} \tag{20}$$

(excuse the occurrence of the letter k in a different sense), Now if the heat-bath is very large, we may set

$$\begin{cases} S(E - E_k) &= S(E) - \left(\dfrac{\partial S}{\partial E}\right)_E \cdot E_k \\ &= S(E) - \dfrac{E_k}{T}, \end{cases} \tag{21}$$

where T denotes the temperature of the heat-bath for the energy E, calculated according to Planck. That is, instead of the ratios (20) we may use the following,

$$\alpha_k e^{-\frac{E_k}{kT}}. \tag{22}$$

Thus we have obtained the important result: The average excitation strengths of the proper values of a system in the heat-bath are proportional to the relative numbers – according to the old quantum statistics – of the members of a canonical aggregate, which occur in the separate states considered

---

prohibition for the proper functions, or when we come to regard certain distributions of energy as essentially different, or not, in Planck's enumeration.

as quantised. The multiplicities of the proper values of the system in question appear as "quantum weights."

We can also get rid of the original assumption that a single proper value E is excited in the *whole* system. This procedure corresponds exactly to that made in classical statistics when we start from a micro-canonical aggregate and prove that a small partial system is distributed canonically in phase. If we wish, however, we can always make a canonical distribution for the whole system in addition; the result for the partial system remains unaltered. Of course the same is true in our case also.

The result (22) should in principle suffice to enable us to transfer all the important results of the old quantum statistics, in particular the statistical theory of gases, of solids, and of the "hohlraum" (Planck's radiation formula) – since all can be based on this formula – into the new theory without difficulty; of course the larger or smaller alterations alluded to in the footnote on p. 227 must be made. I should like to lay special emphasis on the fact that this transference is *possible*, even *without* the support of the quantum postulates.

If the reader likes, however, he can understand everything that has been said in this paper in accordance with Born's theory,[12] in which the postulates are retained and the squares of the amplitudes are not interpreted as simultaneous excitation strengths in the single system, but merely as probabilities (relative frequencies of occurrence) of the discrete quantum states in a virtual aggregate. I have tried to think over the question whether we might from this point of view be able to do without the statistical hypothesis of § 3. This does not seem to be the case. According to Born, the alteration of the "probability field" as time goes on is compulsorily (causally) controlled by the wave equation, and consequently the alteration in time of the "probability amplitudes" is controlled by

---

[12]M. Born, *Ztschr. f. Phys.* 37, p. 863; 38, p. 803; 40, p. 167, 1926.

the equations (9). Hence the objection to reversal mentioned in § 3 now applies to the alteration in time of the probability amplitudes. So far as I can see, we can therefore never reach a one-way (irreversible) course without a supplementary hypothesis about the relative probability of the various possible distributions of the initial values of the probability amplitudes. I am averse to this conception, not so much on account of its complexity as on account of the fact that a theory which demands our assent to an absolute primary probability as a law of Nature should at least repay us by freeing us from the old "ergodic difficulties" and enabling us to understand the one-way course of natural processes without further supplementary assumptions.

Zürich, Physical Institute of the University.
(Received June 10, 1927.)

www.ingramcontent.com/pod-product-compliance
Lightning Source LLC
Chambersburg PA
CBHW021921190326
41519CB00009B/867